Land Policies in Europe

Thomas Hartmann · Andreas Hengstermann ·
Mathias Jehling · Arthur Schindelegger ·
Fabian Wenner
Editors

Land Policies in Europe

Land-Use Planning, Property Rights,
and Spatial Development

 Springer

Editors
Thomas Hartmann
Department of Spatial Planning
TU Dortmund University
Dortmund, Germany

Mathias Jehling
Leibniz Institute of Ecological Urban
and Regional Development
Dresden, Sachsen, Germany

Fabian Wenner
RheinMain University of Applied Sciences
Wiesbaden, Hessen, Germany

Andreas Hengstermann
Department for Urban and Regional
Planning
Norwegian University of Life Sciences
Ås, Norway

Arthur Schindelegger
BOKU University
Vienna, Austria

ISBN 978-3-031-83724-1   ISBN 978-3-031-83725-8 (eBook)
https://doi.org/10.1007/978-3-031-83725-8

This Springer imprint is published by the registered company Springer Nature Switzerland AG
The registered company address is: Gewerbestrasse 11, 6330 Cham, Switzerland

If disposing of this product, please recycle the paper.

Foreword–30 years of land policy research in Europe

Abstract It is very pleasant that this publication provides an up-to-date international comparison and discussion of land policy approaches in Europe, so that we can learn from each other. In recent years sustainable and more socially acceptable approaches to land policy are required. They are now on the political agenda in many European countries and are demanded by the general public. Therefore, a scientific publication that presents experiences from 12 European countries following a mutual case-based approach is beneficial.

Contemporary challenges of land policy

Land policy challenges are increasingly linked to various objectives. At the top of the agenda (once again) is the inability of land and property markets to meet the needs of people to satisfy the basic need for housing. In many places – especially in cities and tourist centres – the desired socially appropriate housing supply for large parts of the population is not (or no longer) guaranteed due to excessively high land and property prices. Irrespective of cyclical market fluctuations, this also applies to the moderate phases of the recurring 'hog cycle'. Building land is essentially available, but in many major European cities its share of the total development costs exceeds 50%. Enormous increases in land values exacerbate the land policy challenges due to speculative behaviour on the land markets. The "market-compliant" solution would require a significant increase in supply of building land and new buildings.

However, this is diametrically opposed to another challenge: The European Union, as part of her sustainable development strategy, pursues the crucial ecological objective of reducing the amount of new land becoming sealed to zero by 2050 (zero net land take), i.e. to freeze the total supply of building land. How can these two challenges and goals be reconciled? Or is it an insoluble dilemma – at least at the level of a single municipality? Solutions are often discussed, but conclusive solutions and effective monitoring are still not in sight.

The third key challenge for policymakers concerning the "land question" arises from the land requirements for climate change mitigation and adaptation, e.g., activating land for flood retention or renewable energies. Changing and adapting land use is a complex challenge, not only for planning, but also for implementation. Practical solutions that provide sufficient quantities and suitable quality of land in line with climate change requirements can hardly be expected without equalising benefits and burdens between stakeholders involved, especially landowners. Therefore, research on corresponding systems is an increasingly important land policy task on the agenda.

So, land policy challenges have increased in the last decades. The range of topics has expanded, mainly due to the climate change and energy transition. However, some challenges regularly reappear on the political agenda in many countries – such as the recurring problems on the housing market. Sustainable land policy is also being discussed as an essential social component for the necessary transformation in terms of sufficient urban development (Petersen et al. 2023). In some countries, leading associations and social organisations joined forces, e. g. the "Bündnis Bodenwende" (Alliance for a Transition in Land Policy) in Germany was founded in 2020 to strengthen the importance of a sustainable and public welfare-oriented land policy in politics and the public debate (DASL 2021).

Definition of land policy

Land policy is interpreted differently in the countries of Europe, and even more diverse on a global scale, e.g. according to the land problems in the Global South. In a broad understanding of land policy, one might consider the actions of private owners of large estates as land policy, for example, when private individuals or companies act strategically with their land assets and optimise them according to their objectives. Here, however, land policy is understood prior as a social task. It is primarily understood as a strategy of the state and its regional authorities, related to a public concern oriented towards the common good. From an operational perspective, land policy is especially concerned with controlling and influencing the property market in order to secure land policy aims. The objectives of land policy often refer to settlement and infrastructure development as well as property rights. Inherent to this understanding is an appropriate balance between the interests of the common good, which manifest themselves in sovereign planning, and private property rights (Davy 2012). The derived land policy strategies in the countries range from "planning-dominated approaches" (plan-led systems) to "marketdominated approaches" (market-led systems).

In the above sense, a successful and sustainable land policy proves to be particularly complex and multi-layered. Land policy not only has to deal with conflicts of objectives that are difficult to resolve (e.g. expanding supply to stabilise prices versus limiting land use to protect resources) but is also a highly interdisciplinary field. Planning, building law, tax law, fiscal assets and property formation, ecological, social, agricultural, and energy supply-related aspects must be combined to achieve an overall satisfactory result; some of these can be influenced by land use planning and management, while others, such as financial and social aspects, cannot (OECD 2017b). Therefore, effective land policy also depends on appropriate governance and must be a standard and permanent topic on the political agenda, which should be organised in close cooperation with urban and spatial development as well as real estate management. Many studies have shown that a cross-departmental working group or a clear designation of the task for a lead department are essential for a successful implementation of land policy objectives (Pätzold et al., 2023).

Land policy perspectives by comparative studies

The search for solutions, i.e. suitable instruments (frameworks) and pathways (processes), is at the heart of the discussions, as the working group on which this publication is based on has also shown. The publication of research findings and good practice cases is a proven basis and practice for sharing experiences and progress in land policy. In comparative research on land policy, various focal points of knowledge can be identified in recent decades (see e. g. Gerber et al. 2018).

Planning systems are one important framework of land policy. International comparative studies on the structure of planning systems have been repeatedly presented at specific intervals (CEC 1997, Reimer et al. 2014, OECD 2017a, Nadin et al. 2018). From a land policy perspective, however, it is not the formal structure as such but the binding nature of the planning system that is crucial, as this is strongly correlated with corresponding increases in land values. Suppose, the stipulations in the urban development plans are binding and must be complied with for planning permission. In that case, these plans have a much more robust and earlier influence than in a planning system in which the land use decision on the type and intensity of use only becomes binding with individual planning permissions.

The process of implementing plans and aims is another important objective of land policy. On the one hand, the process depends on the interaction of the involved private and public stakeholders which – on the other hand – is impacted by the usage of legal tools (e. g. land banking, pre-emption rights, reallocation of plots, public value capture, tax incentives etc.). Often typical constellations of tools and stakeholders can be seen in a country or region. They can be analysed best and made transparent by case-studies. The focus should be on the intensity of intervention in private property rights and the distribution of development profits. Both are central concerns of land policy (Gerber et al. 2018).

Comparative studies are available for different tools of implementing urban development projects, such as expropriation, pre-emption rights, public land banking or the procurement of land for social housing (Needham 2000); a current overview of public value capture tools in 29 European countries (Halleux et. al. 2023) shows that their use today is – despite all differences between countries – the rule in order to achieve steering effects in land policy.

Taking a European research perspective

To address the above mentioned challenges, a European perspective is crucial. Fortunately, the international working group on "Innovative Land Policy in Europe", organised in the ARL – Academy for Territorial Development in the Leibniz Association (ARL) – and consisting of experts from 12 EU countries, takes these challenges as their starting point. The work follows a comparative tradition, which the authors of this foreword highly appreciate, as they were significantly involved in an early international study on land policies in five countries around 30 years ago (Dransfeld and Voß 1993). The project "EuProMa - European Land and Property Markets" compared planning systems and building land development paths in five leading economies. EuProMa was launched and funded by the German Federal Ministry of Regional Planning, Building and Urban Development in the context of a possible harmonisation of the then-forthcoming European Single Market. The country reports were published in 1993/4 in the book series "European Urban Land & Property Markets" by UCL Press London with the editors Hartmut Dieterich (TU Dortmund), Dick Williams and Barry Wood (University of Newcastle upon Tyne) (Dieterich et al. 1993).

The EuProMa project, which also included building land provision in connection with the stakeholder groups through case studies, continues to represent a certain peculiarity due to the level of detail and breadth. While the country strategies of that time have of course changed, we see the relevance of this research in its methodological nature. The work of the international EuProMa research group was based on a unique examination pattern; structuring of the land policy objectives into "the framework", "the process" and "the outcome"; defining typical land development processes and their use in the involved countries (Dransfeld 2023). Concerning the countries involved, however, recourse to the book series of that time offers the opportunity to analyse the further development of land policy strategies and instruments over a more extended period of time.

From a research perspective, it should be noted that a coherent theory of the land and real estate market that does justice to all aspects is still widely lacking (Kötter et al. 2015, 136). Innovative new approaches – such as those presented in this publication – are in great demand, both from those responsible for land policy at the various levels of government and from the perspective of fundamental and application-oriented research. In addition to analysing individual instruments only, it is advisable to include their systemic environment and to assess them in relation to the results on the market (outcome) wherever possible. An approach based on comparisons of case-studies is ideal for this. Further well-founded land policy research and transfer strategies are essential to overcome the challenges.

Continuity and change

Over the last 30 years, the options and methods for providing building land have not changed fundamentally, but in some countries, different approaches are preferred today. These cases are fascinating to examine in depth. Examples include the Netherlands, where the successful approach of regular early municipal interim acquisition of all development sites has no longer been practised in many municipalities since the mid-1990s, and Germany, where the urban development contract and its expanded possibilities for sharing benefits and burdens between municipalities and private actors have been experiencing a triumphant advance for around 25 years. From the point of view of both research and application, it is still desirable that land policy strategies consisting of several instruments are also increasingly investigated.

Basic research - preferably in international comparative studies – makes it possible to assess one's national situation in a broader context and broadens the range of possible problem-solving proposals. This anthology provides a current overview of the land policy debates in Europe and, at the same time, identifies important starting points for further research. The international working group led by Thomas Hartmann and Andreas Hengstermann contributed excellently to this debate.

Winrich Voß
Leibniz University
Hannover, Germany
voss@gih.uni-hannover.de

Egbert Dransfeld
Institut für Bodenmanagement
Dortmund, Germany
e.dransfeld@iboma.de

References

CEC – Commission of the European Communities (2004). EU Guidelines to support land policy design and reform processes in developing countries. [SEC(2004)1289], Brussels. Retrieved October 9, from https://eur-lex.europa.eu/LexUriServ/LexUriServ.do?uri=COM:2004:0686: FIN:EN:PDF

DASL e. V. – German Academy for Urban and Regional Planning (2021). Bündnis Bodenwende - Soil policy election touchstones for the 2021 federal election. Retrieved October 9, from https://dasl.de/wp-content/uploads/2018/11/Anlage-1_Buendnis-Bodenwende_ Wahlpruefsteine-2021.pdf

Davy, B. (2012). Land Policy: Planning and the spatial consequences of property. Routledge

Dieterich, H., Williams, R. H., & Wood, B. D. (1993). Book series "European Urban Land and Property Markets". Vol. 1-6 (Netherlands, Germany, France, UK, Italy, Sweden). UCL Press

Dransfeld, E. (2023). Kommunales Landmanagement als Voraussetzung für eine gemeinwohlorientierte Wohnungsbaupolitik. Zeitschrift für Geodäsie, Geoinformation und Landmanagement, 6/2023, 342–353. https://doi.org/10.12902/zfv-0449-2023

Dransfeld, E., & Voß, W. (1993). Funktionsweise städtischer Bodenmärkte in Mitgliedstaaten der Europäischen Gemeinschaft – Ein Systemvergleich. Schriftenreihe des Bundesministeriums für Raumordnung, Bauwesen und Städtebau

Gerber, J.-D., Hartmann, T., & Hengstermann, A. (2018). Instruments of Land Policy - Dealing with Scarcity of Land. Routledge

Halleux, J.-M., Hendricks, A., Nordahl, B., & Maliene, V. (2023). Public Value Capture of Increasing Property Values across Europe. vdf Hochschulverlag AG

Kötter, T., Berend, L., Drees, A., Kropp, S., Linke, H. J., Lorig, A., Reuter, F., Thiemann,K. -H., Voß, W., & Weitkamp, A. (2015). Land and real estate management - terms, fields of action and strategies. Journal of Geodesy, Geoinformation and Land Management (ZfV), 3/2015, 136 - 146. https://doi.org/10.12902/zfv-0064-2015

Nadin, V., Fernandez Maldonado, A. M., Zonneveld, W., Stead, D., Dabrowski, M., Piskorek, K., Sarkar, A., Schmitt, P., Smas, L., Cotella, G., Janin Rivolin, U., Solly, A., Berisha, E., Pede, E., Seardo, B. M., Komornicki, T., Goch, K., Bednarek-Szczepańska, M., Degórska, B., ... & Münter, A. (2018). COMPASS – Comparative Analysis of Territorial Governance and Spatial Planning Systems in Europe. Applied Research 2016–2018: Final Report. ESPON & TU Delft

Needham, B., & de Kam, G. (2000). Land for Social Housing. CECODHAS

OECD - Organization for Economic Co-operation and Development (2017a). Land-Use Planning Systems in the OECD. Country Fact Sheets. OECD Publishing. https://doi.org/10.1787/978926 4268579-en

OECD - Organization for Economic Co-operation and Development (2017b). The governance of land use - Policy Highlights. www.oecd.org/cfe/regionaldevelopment/governance-of-land-use-policy-highlights.pdf

Pätzold, R., Frölich von Bodelschwingh, F., & Bunzel, A. (2023). Practice of municipal building land mobilization and land policy. Difu Impulse 3/2023. German Institute of Urban Affairs (Difu). https://doi.org/10.34744/difu-impulse_2023-3

Petersen, D. J., Christ, M., & Carstensen, J. (2023). Genug Stadt Krisen. Mit Suffizienz und nach-haltiger Bodenpolitik für lebenswerte Kommunen sorgen. Europa-Universität Flensburg / City of Flensburg. Retrieved October 9, from https://www.uni-flensburg.de/nec/nec-news/news/han dreichung-genug-stadt-krisen

Reimer, M., Getimis, P., & Blotevogel, H. (2014). Spatial Planning Systems and Practices in Europe: A Comparative Perspective on Continuity and Changes. Routledge

Winrich Voß   was a Professor of Land and Property Management at Leibniz Univer-sität Hannover from 2006. Since 2023, he has been professor emeritus. His research and teaching cover contributions to spatial development with a focus on land policy, project implementation and conversion strategies in urban and rural areas. After studying Geodesy (RWTH Aachen / University of Bonn), professional as well as academic practices followed. He received his doctorate from the Faculty of Spatial Planning at the University of Dortmund in 1993 (subject: comparison of land markets in five European countries). Since 2006, he has been engaged in different national and international academic organisations in his field and as a Spokesperson for the inter-disciplinary Leibniz Research Centre TRUST (Spatial Transformation – the Future for Urban and Rural Areas).

Egbert Dransfeld   is director and owner of the Institute for Landmanagement, a consulting business in Dortmund, Germany. He advices municipalities, other public institutions and private owners with regard to land policy strategies, land develop-ment procedures and land and property valuations in the context of urban devel-opment projects. He studied Spatial Planning at Dortmund University and received his doctorate in 1993 on the comparison of land markets in Western Europe. 2019 to 2021 he deputized the chair for Land Policy, Land Development and Municipal Surveying, Faculty of Spatial Planning, at the University of Dortmund.

Acknowledgements This book is the result of a significant collaborative effort, and we extend our heartfelt thanks to the many individuals and institutions that supported us as an editorial team. We invite you to take a moment to learn about the various ways in which we received support. The level of support underscores the importance and timeliness of the topic, as recognized by both academic and practice partners.

The book is the culmination of the International Working Group *Land Policies in Europe*, which operated under the auspices of the German Academy for Territorial Development in the Leibniz Association (ARL) from 2019 to 2024. The working group comprised 17 members from 12 countries across Europe, an advisory committee of four members, and received professional, technical, and financial support from the ARL headquarters.

The working group adopted a transdisciplinary approach, actively involving stakeholders from practice in the group meetings. We are particularly grateful to the Metropolitan Region of Rhein-Neckar, the German Federal Ministry of the Interior, Building and Community (BMI), the Association of German Cities (*Deutscher Städtetag*), the Institute for Federal Real Estate (BImA), the City of Munich, and many others who supported us along the way. The insights and discussions with land policy practitioners enriched and inspired the working group.

We deeply appreciate the constructive, friendly, and enthusiastic contributions of all members of the international working group, each of whom brought a unique perspective to the book project. Special thanks go to Brendan Eisenhut and Katharina Künzel for their organizational and editorial assistance, Caspar Kleiner for mapping the case studies, and Sophie Weidig and Uwe Grützner for their creative input on the cover page. Ayla von Essen's meticulous proofreading was also instrumental in ensuring the quality of the final product. We also owe a debt of gratitude to our advisory board members, Rachelle Alterman, Ben Davy, Tejo Spit, and Stéphane Nahrath for their invaluable guidance. Unfortunately, Ben Davy passed away a few weeks before the publication of this volume. We will remember him as an important academic who supported this book project a lot.

We would like to thank all our hosts, who did a fantastic job organizing the Working Group Meetings in Mannheim, Freinsheim, Dortmund, Munich, Ås, and Vienna. These meetings enabled us to develop our topics in inspiring surroundings.

We also extend our gratitude to the International Academic Association on Planning, Law, and Property Rights (PLPR) for enabling us to meet regularly as part of their annual conference in Ústí nad Labem, Ghent, Ann Arbor, Munich, and Cardiff. Their cooperation helped us maintain the working group's dialogue during the pandemic by utilizing their digital platforms.

Most importantly, this project came to life with the generous and tremendous support from the German Academy for Territorial Development in the Leibniz Association (ARL). We received great support from colleagues at the ARL headquarters, including Sebastian Krätzig, Martin Sondermann, Tanja Mölders, and Evelyn Gusted, as well as from the supporting staff, Kathrin Kube, Vanessa Mena, Tanja Ernst, and many others. Andreas Klee, the secretary-general of ARL at the time, was always supportive of our international working group's activities. This is particularly remarkable as the ARL allowed us to complement their thematic scope on spatial and regional planning with the scale of urban development and property.

We are immensely grateful for the collective effort that made this book possible and hope that it serves as a valuable resource for future research and practice in land policies.

Thomas Hartmann
Andreas Hengstermann
Mathias Jehling
Arthur Schindelegger
Fabian Wenner

Competing Interest The book results from a collaboration of the editors and contributors within the International Working Group '*Land Policies in Europe*', a network that has been funded and facilitated by the German Academy for Territorial Development in the Leibniz Association (ARL) from 2019 to 2024. ARL also funded the open access fee for this publication.

Figures 2.1, 3.1, 3.2, 3.3, 4.1, 5.1, 6.1, 7.1, 8.1, 9.1, 10.1, 11.1, 12.1 and 13.1 were created using ArcGIS® software by Esri. ArcGIS® and ArcMap™ are the intellectual property of Esri and are used herein under license. Copyright © Esri. All rights reserved. For more information about Esri® software, please visit https://www.esri.com.

Contents

Reflections

Opening

Introducing Land Policies in Europe

Thomas Hartmann, Andreas Hengstermann, Mathias Jehling, Arthur Schindelegger, and Fabian Wenner

1 What is Land Policy?

Defining land policy universally is a complex challenge due to the diverse perspectives on the topic. Gerber et al., (2018) argue that land policy interprets the relationship between landowners and public interests, aiming to prevent or resolve conflicts over land use. It establishes rules for who can use land and how (Needham, 2006). Land policies involve public interventions in the allocation and distribution of land (Needham et al., 2018) and in the land market (Davy, 2005; Dieterich et al., 1993; Dransfeld & Voß, 1993). This definition suggests that land policies are fundamentally public activities, that require democratic legitimacy, following the rule of law. Additionally, it implies that land policies pursue politically defined social, economic, or ecological objectives. Given their political nature, the effectiveness of land policies must be measured against these objectives. The goals of land policies impact both the efficient allocation of land as a scarce resource and the just distribution of its costs

T. Hartmann (✉)
TU Dortmund University, Dortmund, Germany
e-mail: thomas.hartmann@tu-dortmund.de

A. Hengstermann
Norwegian University of Life Sciences, Ås, Norway
e-mail: andreas.hengstermann@nmbu.no

M. Jehling
Leibniz Institute of Ecological Urban and Regional Development, Dresden, Germany
e-mail: m.jehling@ioer.de

A. Schindelegger
University of Natural Resources and Life Sciences, Vienna, Austria
e-mail: arthur.schindelegger@boku.ac.at

F. Wenner
RheinMain University of Applied Sciences, Wiesbaden, Germany
e-mail: fabian.wenner@hs-rm.de

© The Author(s) 2025
T. Hartmann et al. (eds.), *Land Policies in Europe*,
https://doi.org/10.1007/978-3-031-83725-8_1

Table 1 "Land policy" in different national languages

Country	Term
Austria	*Bodenpolitik*
Belgium	*Grondbeleid [Flemish]* *Politique foncière [French]*
Czechia	*Pozemková politika*
England	*Land Policy [term not established]*
Finland	*Maapolitiikka*
France	*Politiques foncières*
Germany	*Bodenpolitik*
Norway	*Arealpolitikk*
Poland	*Polityka gruntowa*
Sweden	*Markpolitik*
Switzerland	*Bodenpolitik [German]* *Politique foncière [French]* *Politica del suolo [Italian]*
The Netherlands	*Grondbeleid*

and benefits (Needham et al., 2018). A commonality among these broad definitions is that land policies engage with property rights.

In many countries represented in our book, land policy is not recognised as an independent formalised field of public policy. Instead, it is embedded in various domains, such as spatial and land use planning, agricultural law, tax law, and environmental law. Moreover, interpretations of the term 'land policy' differ across Europe. A European debate on land policy has emerged only recently, especially when compared to national discussions. This variation can be observed in the differing translations of the term used in national languages (see Table 1).

On a more operational and practical level, in many of the countries represented here, land policies are understood as specific strategies within urban development that utilise policy instruments to deliberately intervene in property rights. The Dutch *grondbeleid*, often referred to as active land policy, has garnered significant attention in academic literature and is central to the planning debate. It involves local, public authorities actively participating in the land market to steer spatial development and meet urban development objectives by purchasing and developing land, leveraging the advantages associated with landownership. Similarly, in Sweden and Switzerland, the public sector plays an active role in the land market using civil law instruments, referred to as *Markpolitik, Bodenpolitik, Politique foncière,* or *Politica del suolo*—primarily aimed at providing affordable housing.

In Germany and Austria, land policy is referred to as *Bodenpolitik*, but it encompasses a much broader interpretation, including national legislation and municipal urban development strategies as well as technical land management. In France, the term *politiques foncières* is plural, reflecting the multiple understandings and functions across various policy fields that contribute to denser construction or reduce land consumption (see also Davy's reflection in this book). In some countries, land policy is nearly synonymously with spatial planning. For example, Norway's *Arealpolitikk* represents contemporary spatial planning that emphasises sustainability through densification. Conversely, Finland's *Maapolitiikka* is a distinct discipline that collaborates with spatial planning while maintaining legal and political independence.

In countries like Czechia and Poland, land policy strategies exist, but are not as explicitly labelled. Literally translated, it means *Pozemková politika* or *Polityka gruntowa*. This variation underscores the need for discussions of land policies in Europe to embrace the diverse meanings and interpretations, fostering a common ground for comparative studies. For the purposes of this book, we have developed a common, narrower working definition at an operational level: Land policy is the sum of intended interventions that implement land use planning by aligning use and disposal rights in land through strategic combinations of land management instruments. In the conclusion, we will revisit the question of how this broad understanding can be further specified based on the cases presented.

2 Land and Land Policy

Land policies, regardless of varying interpretations, address the inherently scarce resource of land. Society has many needs that are expressed through current and potential land use as well as through control over landowner rights. Over time, political discussions in land-related fields lead to the evolution of land policies, shaped by shifting priorities. Presently, issues such as housing affordability, land take, and urban densification are widely debated and are thus also reflected in this book's case studies.

Many European countries face a significant shortage of affordable housing in rapidly growing urban areas. In response, several nations have established quantitative housing targets– such as the Netherlands' goal of 650,000 new dwellings by 2040 and Germany's target of 400,000 new apartments annually. Various factors contribute to this shortage, including inter-regional migration and increasing demand for living space (Haase et al., 2013). While the specific objectives vary, (for instance, in terms of affordable housing), the general issue of housing scarcity is relatively common across Europe (Ehrhardt et al., 2022; Fields & Hodkinson, 2018; Wetzstein, 2017). Different countries implement distinct public policies to address these challenges.

Simultaneously, the political debate around reducing land take—the conversion of natural or agricultural land for urban or infrastructure development—has gained prominence in Europe. The European Commission has set a political goal of achieving no net land take by 2050 (Schatz et al., 2021), with similar national targets established in some countries. The United Nations' Sustainable Development Goal 11 also emphasises that global urban sprawl poses significant risks to creating inclusive, safe, resilient, and sustainable cities (Behnisch et al., 2022). Europe's perception of land scarcity stems from factors such as intensive land use (Marquard et al., 2020), an oversupply of building land for urban expansion in recent decades (Lacoere et al., 2023), but also culturally engrained views on land use.

As a result of the competing policy objectives, reducing land take while addressing housing shortages, the policy of urban densification emerged (Dembski et al., 2020). Densification involves increasing the number of inhabitants or built-up space in existing urban areas, and can take various forms. This includes infill development in residential areas, the transformation of brown- and greyfields, or the use of remaining urban green spaces. Soft densification refers to reproposing offices, shops, or attics into residential use, subdividing apartments or expanding existing buildings (Götze & Jehling, 2023). Urban densification—with its specific demands for land within developed areas and the need for mixed land use—heavily relies on land policy and the management of complex property rights (Hartmann et al., 2023).

These policy challenges and objectives are interconnected, with each influencing the other: Urban development, the protection of natural areas, and the preservation of agricultural land often compete for land resources. Urban (re)densification is increasingly seen as a potential political compromise. To explore how different countries navigate such conflicts over land use, this book examines common land use conflicts across various European countries.

3   Actors of Land Policy

The landowner is undeniably the most crucial actor in spatial planning. Recent planning challenges, such as the aforementioned housing shortage, land use reduction, and densification, highlight the significance of the relationship between planners—understood as institutional actors representing institutional rationales—and landowners, who ultimately control development (Hengstermann et al., 2023). Despite this, planning practitioners often perceive landowner as obstacles to implementing spatial plans.

However, the roles of these actor groups are not fixed. Three key aspects make a closer examination of these actors essential to understanding land policy:

First, neither group is homogeneous. Private landowners vary in concerning usage and economic interests, including individual owner-occupiers, large-scale real estate companies, or public- and semi-public entities like housing associations and cooperatives. Furthermore, within the land development process, roles may differ—ranging from original landowners to investors, builders, and developers (Meijer & Buitelaar, 2023). Similarly, planners are also a diverse group. Planning authorities typically operate as the executive branches of local, regional, or national governments, which may even compete with one another over policy objectives. This competition can significantly impact their capacity to pursue land policy (van Oosten et al., 2018). Recognising the heterogeneity of these actors is crucial to understanding their strategies and behaviours.

Second, the roles of these actors can overlap, merge, or even shift depending on specific land policies. In some cases, land policies require a clear distribution of roles between private entities and public authorities, such as in expropriation or land readjustment processes. However, in other cases, public actors—such as municipalities—may act as real estate developers while still playing an active role in public decision making (Meijer & Jonkmann, 2020, Hartmann & Spit, 2015). This blurs the boundaries between interests and agency of the actors involved.

Third, each actor brings their own knowledge, experience, and rational to the table. This means that, beyond legal frameworks and rights, actors may not always behave according to the logic planning laws predict. Instead, adopting a neo-institutional perspective (Debrunner et al., 2020), their actions may appear inconsistent or even irrational from an external point of view (Davy, 2014). Understanding these diverse rationalities is critical for land policy, which seeks to influence the behaviour of specific actors' groups.

4   Institutions and Instruments of Land Policy

The relationship between landowners and planning authorities is institutionalised through planning laws and property rights, which together frame the field of land policy. Planning and property rights exist in a "fraught relationship" (Buitelaar, 2012): While definitions of spatial planning vary, most involve considerations of 'the future' and 'space' in a broader or narrower form (Hillier, 2010). Essentially, spatial planning aims to influence what happens on the ground, meaning every spatial planning activity ultimately concerns individual piece of land. In countries with liberal constitutions, these pieces of land are owned by one or more landowners who hold property rights.

Property rights in land and real estate are afforded special protection through constitutional provisions as they are seen as essential to promoting "individual liberty, political stability, and economic prosperity" (Ellickson, 1993, p. 1317). Any intervention in these property rights must be justified by serving a societal goal (Hartmann & Needham, 2012b, p. 4). Spatial planning often proposes such interventions. It encompasses many interrelated plans and programmes across scales, and is executed by different tiers of government—ranging from cross-national regional planning to municipal land use planning, involving both formal and informal approaches, as well as sectoral and comprehensive planning. Both property rights and planning are recognised social constructions (Bromley, 2000; Hartmann & Needham, 2012b; van Straalen et al., 2018). Within these institutions, land policies have a range of instruments at their disposal.

While spatial planning steers development by determining rights of use, land policies concern rights of disposal through various instruments. This creates a blurred distinction between spatial planning and land policy, as planning administrations have numerous tools that combine the allocation of use and distribution of rights, such as land readjustment, land consolidation, expropriation, and urban development contracts (Hengstermann & Hartmann, 2018; Shahab et al., 2021). These instruments are typically embedded in planning or building laws and aim to influence the behaviour of relevant actors—often landowners—whose behaviour is perceived to contribute to collective problems (Gerber et al., 2018).

These instruments can be applied with varying degrees of flexibility by decision-makers, such as municipal councils and regional planning bodies (Gerber et al., 2018). The decision to use a specific instrument is primarily a matter of political priorities and, in practical terms, depends on the legal framework and principles of administrative action, such as proportionality and the rule of law. Practical considerations, like the public authority's budget, may also play a role.

Land policies often strategically combine different regulatory or fiscal instruments to achieve specific planning objectives. These pathways can evolve either inductively based on established practices or deductively from overarching theoretical or political frameworks (Shahab et al., 2021).

The implementation of land policies involves various technical and administrative aspects, informed by disciplines such as law or economics, which describe and predict the behaviour of actors under certain circumstances. For example, classical location theory (Alonso, 1964) is critical for understanding the causes and effects of certain allocations. However, beyond these technical and administrative considerations, land policy is also highly normative. As a form of public policy, it must balance public and private interests in land use. Spatial planning often institutionalises the public interest, while property rights represent private interests.

5 Understanding, Comparing, and Reflecting on Land Policies

This book is entitled "Land Policies in Europe", with the plural 'policies' emphasising both the multitude of implementation pathways within individual countries as well as the differences between countries. The book has three main objectives: to understand, compare, and reflect on land policies in Europe. Understanding land policies involves grasping the concepts, resources, actors, and institutions that shape these policies in each country. The structure of the case studies presented in this book enables exploration and comparison of different land policies, which is essential for reflecting land policies within their respective national contexts.

Planning practitioners often do not fully consider the diverse perspectives of landownership, spatial planning practice, and planning theory concerning the relationship between property rights and spatial planning (Hartmann & Needham, 2012a). However, "while there has been a recent surge in academic interest in the development of new strategies and instruments of land policies, it appears that national discussions rarely inform each other" (Hengstermann et al., 2023). This book aims to provide a structured reflection on land policies across Europe, highlighting the various ways different countries approach the relationship between planning and property rights. Such reflection can break through the impasses of a "property blind" planning, offering new perspectives and approaches (Hartmann & Needham, 2012b;

van Straalen et al., 2018), ultimately leading to more effective, efficient, legitimate, and just land policies (Gerber et al., 2018).

Reflecting on land policies requires stepping outside the national contexts in which they are framed to explore alternative policy-driven solutions to common land-related challenges. Each country socially constructs its own interpretations of planning and property rights. While these interpretations are institutionalised through constitutions and legislation, legitimising complex governance structures across on varying levels of government, the practical interpretation of the relationship between property rights and planning ultimately materialises on individual plots of land. For this reason, the approach of this book begins with individual plots, creating a common ground—both literally and figuratively—for understanding how land policies intervene in property rights to meet spatial planning objectives.

6 Structure of the Book

The book adopts a problem-based approach, starting with a plot of land on which land policies are implemented. Each of the 12 countries is represented through selected case studies that illustrate specific problems and potential solutions in an intuitive and accessible way (see Fig. 1). These case studies offer a narrative perspective aimed at stimulating intellectual engagement. Each chapter concludes with a reflection section, inviting readers to consider how the case studies relate to their own country's land policy. Unsurprisingly, the reflections suggest that while each case includes *some* typical aspects, many others make these case studies unique. It is up to the reader to identify which aspects are most relevant. The cases do not generally represent the country in a statistical-scientific sense. However, the portray typical mechanisms and offer accessible insights for an international audience.

The case studies are presented in a consistent format, comprising five sections.

1. The country's understanding of land policy is presented.
2. This is followed by a description of the problem, conceptualising land as a resource whose sustainability is variably threatened.
3. An in-depth analysis of the actors involved is provided, shedding light on their capabilities, interests, and influence.
4. The institutional framework governing land use within the specific case, as well as its broader context, is outlined.
5. The final section reflects on the extent to which the case presents a prototypical example of the country's land policy.

Authors are encouraged to include additional aspects that may contribute to a broader understanding of land policy even if these are not fully embedded in the chosen case. Rather than taking an encyclopaedic approach, this problem-based presentation of land policies in Europe moves from a small-scale focus on specific cases to a broader understanding of overarching patterns and issues. Each section examines a specific case, subsequently contextualising it within the broader

Fig. 1 Cases in 12 countries (own figure)

framework. The three reflective chapters by Davy, Spit, and Dembski highlight the differences and commonalities across the cases.

This approach invites two likely critiques. First, the book compares seemingly disparate paradigms. The focus on land policy means engaging with understood differently in each country. However, the rationale behind this comparative method is that the contrast reveals nuances that enrich the understanding of land policies across Europe. As readers progress through the chapters, they will observe occasionally comparing concepts that aren't entirely equivalent—as evidenced by the antidictionary sections of every chapter. Second, the selection of countries is inherently limited, and this book cannot feasibly cover every nation and its respective land policies. Instead, it offers an initial exploration with potential for future expansion. Even comprehensive coverage, any study of land policy would remain inherently incomplete due to variations within individual countries. Despite the validity of these critiques, this book makes a substantial contribution to planning theory, stimulating discourse on land policies not only within Europe, but on a global scale.

The book does not advocate directly transplanting successful aspects of one country's land policy to another. Such legal transplants must consider national differences—legal, political, and social—if they are possible at all. Instead, the book aims to encourage readers to reflect on their own country's land policies by being exposing them to what is happening elsewhere, potentially right next door.

By presenting a range of experiences from different countries, the book aims to inspire scholars, practitioners, decision-makers, and, of course, students to become curious about land polices in other nations. The book shows how many countries face similar challenges such as housing shortage, urban sprawl, and densification. These spatial challenges are subject to scientific debate and legislative actions across Europe. The approaches taken in different jurisdictions should serve as inspiration for reflection on one's own context.

Willem Salet, in his review of "Instruments of Land Policy" (Gerber et al., 2018), calls for "systematic and critical attention to the institutional underpinning of the use of policy instruments" (Salet, 2018). This book takes up that challenge, discussing land policies in Europe—not as an encyclopaedic comparison, but as a work that encourages and enables reflections on land policies beyond national borders. The editors and authors developed this book out of a belief that international perspectives on land policy can inspire and enrich reflections on these crucial issues, linking back to the overarching question of land.

7   The Land Question

The land question, a fundamental topic that ties into broader theories of the state, has its roots in the era of industrialisation and urbanisation during the 19th and early twentieth centuries. This period was characterised by phenomena such as housing shortages, overcrowded cities, poor living conditions, and significant "unearned" profits for landowners. At its core, the land question revolves around the issue of who should benefit from the land rent. Several land reformers such as Henry George, Adolf Damaschke, Hans Bernoulli, and Ebenezer Howard, approached this question from different perspectives and proposed varying solutions. Bernoulli, for instance, advocated for a pragmatic yet equitable distribution of land value by separating land ownership from the right to use it. George took a more radical stance, calling for a land value tax. Damaschke emphasised the fair distribution of property rights across society, while Howard combined community decision-making with urban planning. Many other key thinkers and concepts contributed to the debate. Historically, the land question has centred on the distribution of profits from land use and the increase in land value. Ultimately, the land question is a political one.

This political debate largely faded from the agenda in many European and North American countries during the latter half of the twentieth century, as rise of car-dependent mobility and accompanying suburbanisation reduced the perceived scarcity of land—even though the basic mechanisms of land markets remained fundamentally unchanged. In contrast, countries in the Global South continued to grapple with issues such as lack of land tenure security, limited access to ownership, and rampant land grabbing (Banon & Jehling, 2020). In the Global North regulation of land use came to be viewed increasingly as a technical issue—particularly in the field of spatial planning (Baum, 1977; Witte & Hartmann, 2022). A trend of de-politicisation, which dominated the early 2000s, affected discussions surrounding the land question (Wood, 2016). However, land has once again become a politicised issue. Politicisation refers to the process by which an issue shifts from being perceived as a matter of fate or inevitability to one that is subject to debate and contingency (Hay, 2013). This process gives an issue purpose and meaning (Hay, 2013). While the land question historically focused on social inequalities tied to the distribution of land rent, today's contemporary land question adds ecological concerns, especially those related to climate protection and adaptation. Additionally, modern discussions around the land question address political goals aimed at reducing land consumption (Marquard et al., 2020). Thus, the contemporary land question incorporates the issue of the scarcity of land into a broader, social dimension.

The land question is by no means merely an academic debate. It touches on the very foundations of society. Davy (2014) illustrates how property rights links to the ideas of thinkers like Thomas Hobbes, Jean-Jaques Rousseau, and John Locke (Davy, 2014). Similarly, Nikolić (2018) shows how different "models of ownership relations have been cyclically gaining primacy in Europe. One is characterised by legal and economic liberalism, while the other is by a pronounced state interventionism" (Nikolić, 2018). These models vary in how they define the relationship between the state and private landowners. In essence, land policy shapes, reinforces, or redefines the fundamental structures of society, requiring continuous reflection. This reflective process is encouraged by Land Policies in Europe.

References

Alonso, W. (1964). *Location and land use. Publications of the Joint Center for Urban Studies.* Harvard University Press.

Banon, F., & Jehling, M. (2020). Looking for innovation—Trajectories of land transaction and readjustment in West Africa. *Cities, 106*, 102880. https://doi.org/10.1016/j.cities.2020.102880

Baum, H. S. (1977). Toward a post-industrial planning theory. *Policy Sciences, 8*(4), 401–421.

Behnisch, M., Krüger, T., & Jaeger, J. A. G. (2022). Rapid rise in urban sprawl: Global hotspots and trends since 1990. *PLOS Sustainability and Transformation, 1*(11), e0000034. https://doi.org/10.1371/journal.pstr.0000034

Bromley, D. W. (2000). Regulatory takings and land use conflicts. In M. D. Kaplowitz (Ed.), *Property rights, economics, and the environment* (pp. 23–33). JAI Press.

Buitelaar, E. (2012). The fraught relationship between planning and regulation: Land use plans and the conflicts in dealing with uncertainty. In T. Hartmann & B. Needham (Eds.), *Planning by law and property rights reconsidered* (pp. 207–218). Ashgate.

Davy, B. (2005). Bodenpolitik. In E. H. Ritter (Ed.), *Handwörterbuch der Raumordnung* (4., neu bearb, pp. 117–130). ARL.

Davy, B. (2014). Polyrational property: rules for the many uses of land. *International Journal of the Commons, 8*(2), 472. https://doi.org/10.18352/ijc.455

Debrunner, G., Hengstermann, A., & Gerber, J.-D. (2020). The business of densification: Distribution of power, wealth and inequality in Swiss policy making. *Town Planning Review, 91*(3), 259–281. https://doi.org/10.3828/tpr.2020.15

Dembski, S., Hartmann, T., Hengstermann, A., & Dunning, R. (2020). Introduction: Enhancing understanding of strategies of land policy for urban densification. *Town Planning Review, 91*(3), 209–216. https://doi.org/10.3828/tpr.2020.12

Dieterich, H., Williams, R. H., & Wood, B. D. (1993). *Book series "European Urban Land and Property Markets"*. Vol. 1–6 (Netherlands, Germany, France, UK, Italy, Sweden). UCL Press.

Dransfeld, E., & Voß, W. (1993). *Funktionsweise städtischer Bodenmärkte in Mitgliedstaaten der Europäischen Gemeinschaft – Ein Systemvergleich*. Schriftenreihe des Bundesministeriums für Raumordnung, Bauwesen und Städtebau.

Ehrhardt, D., Eichhorn, S., Behnisch, M., Jehling, M., Münter, A., Schünemann, C., & Siedentop, S. (2022). Stadtregionen im Spannungsfeld zwischen Wohnungsfrage und Flächensparen. Trends, Strategien und Lösungsansätze in Kernstädten und ihrem Umland. *Raumforschung Und Raumordnung, 80*(5), 522–541. https://doi.org/10.14512/rur.216

Ellickson, R. (1993). Property in Land. *Yale Law Review* (102), 1315–1400.

Fields, D. J., & Hodkinson, S. N. (2018). Housing Policy in Crisis: An International Perspective. *Housing Policy Debate, 28*(1), 1–5. https://doi.org/10.1080/10511482.2018.1395988

Gerber, J. D., Hartmann, T., & Hengstermann, A. (Eds.). (2018). *Instruments of land policy: Dealing with scarcity of land*. Routledge.

Götze, V., & Jehling, M. (2023). Comparing types and patterns: A context-oriented approach to densification in Switzerland and the Netherlands. *Environment and Planning B: Urban Analytics and City Science, 50(6)*, 1645–1659. https://doi.org/10.1177/23998083221142198

Haase, D., Kabisch, N., & Haase, A. (2013). Endless urban growth? On the mismatch of population, household and urban land area growth and its effects on the urban debate. *PLoS ONE, 8*(6), e66531. https://doi.org/10.1371/journal.pone.0066531

Hartmann, T., Dembski, S., Hengstermann, A., & Dunning, R. (2023). Land for densification: How land policy and property matter. *Town Planning Review, 94*(5), 465–473. https://doi.org/10.3828/tpr.2022.22

Hartmann, T., & Needham, B. (2012a). Introduction: Why reconsider planning by law and property rights? In T. Hartmann & B. Needham (Eds.), *Planning by law and property rights reconsidered* (pp. 1–23). Ashgate.

Hartmann, T., & Needham, B. (Eds.). (2012b). *Planning by law and property rights reconsidered*. Ashgate.

Hartmann, T., & Spit, T. (2015). Dilemmas of involvement in land management—Comparing an active (Dutch) and a passive (German) approach. *Land Use Policy, 42*, 729–737. https://doi.org/10.1016/j.landusepol.2014.10.004

Hay, C. (2013). From politics to politicisation: Defending the indefensible? *Politics, Groups and Identities, 1*(1), 109–112. https://doi.org/10.1080/21565503.2012.760315

Hengstermann, A., & Hartmann, T. (2018). Instruments of land policy: Four types of intervention. In J.-D. Gerber, T. Hartmann, & A. Hengstermann (Eds.), *Instruments of land policy: Dealing with scarcity of land* (pp. 27–32). Routledge.

Hengstermann, A., Wenner, F., Jehling, M., & Hartmann, T. (2023). *Innovative land policies in Europe*. In press.

Hillier, J. (2010). Introduction: Planning at yet another crossroads? In J. Hillier & P. Healey (Eds.), *The Ashgate research companion to planning theory: Conceptual challenges for spatial planning* (pp. 1–34). Ashgate.

Lacoere, P., Hengstermann, A., Jehling, M., & Hartmann, T. (2023). Compensating Downzoning. A comparative analysis of European compensation schemes in the light of net land neutrality. *Planning Theory and Practice*. Advance online publication. https://doi.org/10.1080/14649357.2023.2190152götze

Marquard, E., Bartke, S., Gifreu i Font, J., Humer, A., Jonkman, A., Jürgenson, E., Marot, N., Poelmans, L., Repe, B., Rybski, R., Schröter-Schlaack, C., Sobocká, J., Tophøj Sørensen, M., Vejchodská, E., Yiannakou, A., & Bovet, J. (2020). Land consumption and land take: Enhancing conceptual clarity for evaluating spatial governance in the EU context. *Sustainability, 12*(19), 8269. https://doi.org/10.3390/su12198269

Meijer, R., Buitelaar, E. (2023): What drives developers? Understanding vertical (dis)integration strategies in the land development process. *Land use policy, 131*, 106718. https://doi.org/10.1016/j.landusepol.2023.106718

Meijer, R., & Jonkman, A. (2020). Land-policy instruments for densification: The Dutch quest for control. *Town Planning Review, 91*(3), 239–258. https://doi.org/10.3828/tpr.2020.14

Needham, B. (2006). *Planning, law, and economics: The rules we make for using land*. Routledge.

Needham, B., Buitelaar, E., & Hartmann, T. (2018). *Planning, law and economics: The rules we make for using land* (2nd ed.). Routledge.

Nikolić, D. Ž. (2018). Climate change and property rights changes. In F. van Straalen, T. Hartmann, & J. Sheehan (Eds.), *Property rights and climate change: Land-use under changing environmental conditions* (pp. 52–70). Routledge.

Salet, W. (2018). Book review: Instruments of land policy: Dealing with scarcity of land. *Town Planning Review, 89*(6), 651–656. https://doi.org/10.3828/tpr.2018.44

Schatz, E.-M., Bovet, J., Lieder, S., Schroeter-Schlaack, C., Strunz, S., & Marquard, E. (2021). Land take in environmental assessments: Recent advances and persisting challenges in selected EU countries. *Land Use Policy, 111*, 105730. https://doi.org/10.1016/j.landusepol.2021.105730

Shahab, S., Hartmann, T., & Jonkman, A. (2021). Strategies of municipal land policies: Housing development in Germany, Belgium, and Netherlands. *European Planning Studies, 29*(6), 1132–1150. https://doi.org/10.1080/09654313.2020.1817867

van Oosten, T., Witte, P., & Hartmann, T. (2018). Active land policy in small municipalities in the Netherlands: "We don't do it, unless…." *Land Use Policy, 77*, 829–836. https://doi.org/10.1016/j.landusepol.2017.10.029

van Straalen, F., Hartmann, T., & Sheehan, J. (Eds.). (2018). *Routledge complex real property rights series. Property rights and climate change: Land-use under changing environmental conditions.* Routledge.

Wetzstein, S. (2017). The global urban housing affordability crisis. *Urban Studies, 54*(14), 3159–3177. https://doi.org/10.1177/0042098017711649

Witte, P., & Hartmann, T. (2022). *An introduction to spatial planning in the Netherlands.* Routledge.

Wood, M. (2016). Politicisation, depoliticisation and anti-politics: Towards a multilevel research agenda. *Political Studies Review, 14*(4), 521–533. https://doi.org/10.1111/1478-9302.12074

Thomas Hartmann is the chair of land policy and land management at the School of Spatial Planning, TU Dortmund University, Germany. His research focuses on strategies of municipal land policy, and the relation of flood risk management and property rights. He is the former president of the International Academic Association on Planning, Law, and Property Rights.

Andreas Hengstermann is an Associate Professor with the Department of Urban and Regional Planning (BYREG) at the Norwegian University of Life Sciences (NMBU). He holds a Ph.D. in Geography from the Institute of Geography of the University of Bern (CH). He was educated as a planner (Dipl.-Ing. Raumplanung) at the TU Dortmund University (DE) and the Universidad de Huelva (ES). Furthermore, he did postgraduate studies in public law (DAS Law). His primary research interest lies in understanding the essential role of property rights in shaping spatial development. His research includes comparative studies across different geographical contexts, political systems, and legislations. He current serves as Vice-President International Academic Association on Planning, Law, and Property Rights.

Mathias Jehling is senior researcher at Leibniz Institute of Ecological Urban and Regional Development (IOER) in Dresden, Germany, where he leads the research group on "Urban Structure and Policy". His focus is on geographic information in the planning context. He works and teaches (Technical University of Dresden) on urban form and institutionalist approaches to planning and land policies. He obtained his PhD at Karlsruhe Institute of Technology.

Arthur Schindelegger is a postdoc research fellow at University of Natural Resources and Life Sciences, Vienna (BOKU University), Institute of Landscape Planning. He has studied spatial planning at TU Wien and KTH Stockholm and has worked in the private sector for several years. His research is centred around questions on land policy, planning law and the integration of natural hazards and climate change adaptation in planning processes and procedures. He is also a consultant with the World Bank on resilient urban development.

Fabian Wenner is professor for sustainable urban and transport planning at RheinMain University of Applied Sciences in Wiesbaden, Germany. His research focuses on integrated transport and settlement planning through accessibility, instruments of land policy for sustainable urban development, and digitalisation in urban planning.

Country Cases

Land Policy in Austria: Affordable Housing Against the Background of Limited Land Resources

Arthur Schindelegger and Walter Seher

Abstract Austria is for the most part an Alpine country with very limited land resources available for housing. This scarcity in combination with a dysfunctional land market and a strong interest in development for commercial usage leads to comparatively high land prices. Therefore, the typically urban phenomenon of too little affordable dwellings is a pressing issue in many alpine areas. Land policy is in this context typically understood as a strategy to correct the land market and achieve the overall political goal of enabling young people to stay in their home valleys. Municipalities have to establish approaches combining different land policy instruments to actually be able to provide affordable housing. The presented case takes a close look in the actual understanding and application of land policy instruments in the municipality of Sölden/Tyrol. Land ownership is concentrated there in the hands of local farmer families that have great influence on local decision making. There is also no well-established land market, as there is a very limited supply that is by far exceeded by the demand. The case shows a long-term strategic approach of land purchases and the combination of different land policy instruments to solve immediate housing problems in unfavourable framework conditions.

1 Land Policy in Austria: Understanding and Evolution

In Austria, land policy (*Bodenpolitik*) is not a legally defined term and does not refer to a consistent legal system (Binder et al., 1990). The understanding of land policy in Austria is associated with public sector interventions in private property rights on land to fulfil the public interest. This concept aligns with the working definition for this volume: "land policy is the sum of governmental interventions that influence the rights and obligations of landowners, usually in pursuit of social, economic, or

A. Schindelegger (✉) · W. Seher
BOKU University, Vienna, Austria
e-mail: arthur.schindelegger@boku.ac.at

W. Seher
e-mail: walter.seher@boku.ac.at

© The Author(s) 2025   19
T. Hartmann et al. (eds.), *Land Policies in Europe*,
https://doi.org/10.1007/978-3-031-83725-8_2

ecological political aims". Throughout their evolution, land policy approaches in Austria have shown different priorities.

The origins of land as an issue of public policy trace back to the liberal revolution of 1848, which led to the replacement of feudalism with a land tenure system that granted individual property rights on land to larger parts of society, particularly farmers. As a result, the first governmental interventions on land issues were aimed at the agriculture and forestry sectors addressing the need for land consolidation and land transfer restrictions. After World War I, land reform attempts targeting large landholdings of the former nobility only had a minor effect. Governmental interventions in the market for building land started with approaches to regulate speculative urban building activity in the late nineteenth century. A notable example is the Viennese Building Code (*Wiener Bauordnung*), enacted in 1930, as it allowed for far-reaching interventions in private property rights. The urban housing shortage following World War I led to squatting campaigns inspired by the land reform movement. In response, the City of Vienna launched a large-scaled municipal housing programme in the 1920s and 1930s (Mayer et al., 2020).

Post-World War II, Austria saw the introduction of spatial planning as a public responsibility. Since then, the State Spatial Planning Acts provide regulation on building land uses, particularly housing, commercial, and industrial development. Affordable housing became a land policy issue again in the 1970s, with the enactment of the Land Acquisition Act (*Bodenbeschaffungsgesetz*) in 1974 attempting to facilitate land acquisition for social housing demand. However, due to a widely stagnating urban population and an increasing economic orientation towards neoliberal principles, land use in connection with affordable housing disappeared from the political discourse for some decades (Mayer, 2020). In the years to follow, the term "land policy" became affiliated with spatial planning (ÖROK, 1993; ÖROK, 1995). In this context, land policy refers to instruments that influence the institutional organisation of property rights on land in order to implement planning objectives and address planning issues (van Straalen, 2014). In Austria, these instruments are first and foremost deployed to mobilise unused building land, increasing land use efficiency and effectively fighting urban sprawl (ÖROK, 1993). Since the 1990's mobilising building land has been a land policy issue, reflected both in planning legislation and planning practice.

The recent comeback of land policy in Austria is on the one hand fuelled by a broader awareness of the problems connected to land take, such as farmland and biodiversity losses and the various consequences of soil sealing. On the other hand, rapidly rising land prices in the aftermath of the financial crisis, together with an ongoing population growth in cities and peri-urban areas, brought affordable housing back on the political agenda. Both resulted in an intensified focus on land policy in the political and public discussion (among others ÖROK, 2014; Mayer et al., 2020).

Due to these changing priorities and its diverse application, there is a variety of legal provisions referring to the term land policy. Thus, land policy can be understood as an umbrella term which comprises various land-related legislations, with a particular affiliation to spatial planning.

In Austria, land-related legislation is not the result of a consistent objective or a uniform social or political consensus, but rather a product of the political balance of power which comes with regulating different ideas (Binder et al., 1990). Besides the weak constitutional backing for land policy instruments, the scope of existing regulatory approaches is not fully utilised. This particularly applies to a so-far missing implementation of the Land Acquisition Act, for property-related levies and taxes as well as intervention-intensive instruments in spatial planning, such as downzoning (ÖROK, 1993; ÖROK, 1995).

This reluctance to apply regulatory instruments together with a strong constitutional protection of property rights preserves existing ownership structures as well as usage rights, empowering landowners in planning processes. This is especially true for Alpine regions with their topographically limited land availability. Thus, Austrian planning authorities, here in the first place municipalities, rely on persuasive instruments to allocate land for achieving planning goals. In this context, municipalities find a variety of individual and innovative tailor-made solutions.

2 Understanding Land-Related Challenges in Alpine Areas

Austria is well known for its Alpine landscape which covers a majority of the country. In total, 29% of the European Alps are located in Austria, which cover nearly 63% of Austria's total area (Alpine Convention, 2016). These Alpine regions have very limited land resources suitable for urban development and agricultural practices compared to the hilly landscapes in the east of the country which allow for large scale agriculture and vast urban expansion. Therefore, land related challenges differ strongly depending on the topographic conditions. Additionally, extensive natural hazards further limit the potential to develop the Alpine valleys. This leads to a paradoxical situation in which many Alpine municipalities have large territories but hardly any land resources suitable for development. Furthermore, many Alpine municipalities economically rely solely on tourism due to the absence of other options (Bundesministerium für Arbeit & Wirtschaft, 2023).

Former farming villages are now among the top destinations regarding overnight stays and high turnover. This fuels conflicting claims over land use, as the revenue from commercial usage far exceeds those from renting residential dwellings. Another essential factor influencing planning decisions is the availability of land for development. Land transfer of agricultural land favours farmers (pre-emption) and is restricted in order to preserve fertile soils and impede land speculation. Based on the Land Transfer Acts (*Grundverkehrsgesetze*), property transfers of agricultural and forested land are subject to approval by a land transfer authority. For example, the Tyrolean Land Transfer Act provides an exemption from the transfer approval requirement for municipalities and the Tyrolean Land Fund (see Sect. 3.2). For municipalities, this exemption applies if the acquisition is required to fulfil public tasks and if the property in question is located in the municipal territory. In other Austrian states, municipalities or housing associations are substantially limited in their possibilities to strategically purchase land not zoned as building land.

Overall, many economically wealthy but topographically limited Alpine municipalities share the following land-related challenges:

- Land use conflicts: limited zonable space results in a scarcity of land available for housing.
- Conflicting claims: touristic development (commercial accommodation, secondary homes, etc.) limits land for housing and social infrastructure even further and drives up prices.
- Role of property rights: land transfer regulation restricts accessibility to land for municipalities and non-farmers.

The municipality of Sölden exemplifies the land related challenges defined in the previous section. Sölden is internationally well-known for its ski resort. With a nearly 470 km² land area, it is larger than the capital city of Vienna, but only around 1.2% of its territory is suitable for potential urban development (ÖROK-Atlas, 2015). The former farming village has grown to around 3,000 year-round inhabitants while also providing 15,000 beds in accommodation for their 2 million overnight stays per year. Tourism spells out both wealth and doom for the municipality. Developable land is scarce and, besides natural hazards, the achievable revenue boosts land prices. Land zoned for residential purposes does not hold any obligation to fulfil the intended usage (Lienbacher, 2016). The provision of an affordable housing stock for locals without property is therefore a major challenge for the municipality. In this regard, a multi-storey building project was recently developed in the hamlet of Kaisers.

Kaisers is a historic settlement structure within the municipality of Sölden, located in one of the relatively few safe areas for development due to the threat of avalanches, rock fall, and flooding (see Fig. 1). The meadows south of the historic hamlet have traditionally been considered too dangerous because of the potential rockfall from the steep slopes above. However, capitalising on its favourable financial situation, the municipality began purchasing land in this area. The goal was to explore the possibility of developing additional subsidised housing through a housing association, despite the uncertainty of implementing a rockfall protection project. The existing plot and ownership structures in the area were unfavourable for development. To address this, the municipality began preliminary discussions with local landowners to assess their willingness to participate in a land readjustment scheme. This scheme aimed to reorganise the plots to facilitate development and allocate adequate space for access roads. As a result of these initial talks, a specific area was demarcated, and the readjustment procedure was initiated.

For the municipality to reach an agreement with the landowners, they requested the provision of land to build a rockfall protection dam (hatched area in Fig. 2), the ability to develop infrastructure, and the sale of a certain share of land from every participating landowner in order to create a large enough plot for the housing project. All participating landowners voluntarily agreed to the readjustment scheme, which meant the loss of their right to appeal. This accelerated the process, allowing the municipality to put out a tender for a competition. The winning proposal served as

Fig. 1 Kaisers in Sölden, Austria (map: Kleiner, Jehling 2024, Aerial imagery provider © Esri, Land Tirol—data.gov.at, Maxar, Microsoft)

the basis for amending the zoning plan and issuing a detailed development plan. Construction began in 2015, with the first apartments completed and handed over in 2016. The second construction phase followed in 2019, adding 23 rental flats.

3 Actors in Austrian Land Policy

There is no such thing as one representative land policy in Austria as problems and solutions are both context and actor dependent. At the same time, the Kaisers case clearly outlines which actors are relevant in developing specific solutions for a typical planning problem such as the establishment of affordable housing.

before land readjustment   after land readjustment

Fig. 2 Land readjustment in Kaisers, Sölden (Schindelegger, 2018)

3.1 Private Actors and Property Ownership

Due to Austria's liberal constitution, private actors are essential in land markets. Property is well protected and does not legally entail any social obligations (Berka, 2021). In comparison to other European states, ownership records are completely transparent. Besides the ledger, the publicly accessible land register contains all land-related deeds (§ 7 *Grundbuchgesetz* 1955), guaranteeing transparency on ownership and easements.

As one of Austria's constitutional laws states that property is inviolable (Art 5 *Staatsgrundgesetz* 1867), the right to interfere with private property is strictly limited. Any compulsory acquisition in light of public interest requires compensation. The rules for compensation were defined in 1954 and have been legally valid ever since (Kühne, 1982). Additionally, many constitutional supreme court decisions have helped to clarify the framework conditions for compensating the actual and de facto expropriation (e.g. Auer, 2011).

By design, Austria's planning law disregards physical ownership, instead asking for an abstract and idealised planning process with a rationalistic approach. At the same time owners' interests need to be considered when drafting legally binding decisions on land use and development rights (e.g. Constitutional Court VfSlg 13.282/ 1992). Owners are granted such usage rights but typically have no obligation to execute them within a certain time period (Lienbacher, 2016). However, this has changed over the past years as development obligations can be associated with zoning or re-zoning of building land in many Austrian states (Kanonier, 2020).

Interestingly, municipalities and other public entities often mirror the role of private actors as well. They have the same rights to acquire land and make use of it in their own interest. Of course, public entities are liable to carefully handle property and advocate for the interests of the general public (Berka, 2021).

In the Kaisers case, private actors included in the land readjustment scheme own a major share of the parcels and are crucial for the success of the applied model. The municipality of Sölden is first and foremost a private actor, but it also seeks to meet public interests—here, the development of affordable flats. In this example, all legally binding decisions were taken consensually among the participating property owners after extensive negotiations. This is typical for Austria's land policy; solutions to complex planning problems are found, not against the interests of private actors, but rather in close collaboration with them.

3.2   Public Actors and Land Policy

The Kaisers case might give the impression that municipalities are the only public entities relevant in the development of tailor-made land policy processes. However, the role of the states, the federal republic and land funds as public actors can be further elaborated.

Indeed, municipalities oversee spatial planning within their territory and formulate strategic, detailed urban development and zoning plans. Decisions have to be taken by the municipal councils (Kanonier & Schindelegger, 2018). Furthermore, municipalities manage the local roads that are public property and—often together with other municipalities organised in associations—run the local technical and social infrastructure. Municipal responsibilities are therefore manifold, solidifying municipalities as essential players when it comes to urban development.

Typically, the number of different public institutions that need to be consulted and integrated in the development for land policy schemes to establish an affordable housing stock makes the management complex and time consuming. It calls for an experienced and professional institution taking care of the organisational work, which an average municipality can hardly achieve all on its own.

The states take care of the higher-ranked road infrastructure and play an important legislative and administrative role in land policy. Austrian state parliaments enact spatial planning, building, and land transfer laws. The state administration oversees implementation and enforcement. It is also the supervisory authority for municipalities concerning all legally binding planning decisions. In many states it is also the responsible authority for land readjustment schemes. Depending on the state, the administration also manages land transfer procedures. Additionally, the state governments have a spatial planning responsibility towards regional interests. Despite this, the scope of the states is limited by the autonomy of the municipalities (Kanonier & Schindelegger, 2018).

Another semi-public actor which exists in several states are land funds. Land funds are typically owned by the state but organised as commercial companies serving a non-profit purpose. These land funds are instruments of land acquisition and land banking which support municipalities that do not have the necessary financial resources to acquire land on short notice (e.g. Riedmüller & Erhart, 2020). They acquire suitable properties, mainly for housing and commercial development as well as public infrastructure, and make them available to municipalities against payment. Furthermore, they provide subsidies for municipal land acquisition and infrastructure development by subsidising interest rates.

Established in 1994, the Tyrolean Land Fund (*Bodenfonds*), acquires, stockpiles, and sells land for municipal public development. The fund is responsible for operational activities such as concluding contracts and financial processes. Additionally, it manages the land held in reserve including the sale and use of land for exchange purposes. Typically, the Tyrolean Land Fund acquires land which is not yet zoned for development. Being a beneficiary of the Tyrolean Land Transfer Act, the fund is entitled to purchase agricultural and forested land without authorisation of the land transfer authority (Kraus, 2018).

The federal republic is also relevant in land policy matters in many aspects. As a legislator in different sectors (e.g. tax law, forestry law, water law) it sets framework conditions and is in control of constitutional law. The republic owns a significant amount of land, inherited from the last emperor who had land ownership all over the territory. Most of this land consists of forests and waterbodies. In the Kaisers case and similar Alpine municipalities, the federal Forest Engineering Service in Torrent and Avalanche Control (WLV) played an important role in enabling development in an area prone to natural hazards. The WLV was responsible for planning and constructing a rockfall protection dam, which provided the prerequisite for the zoning of building land in this area.

In the Kaisers case study, public actors played an essential role in developing a customised process that generates benefited private actors, encouraging their involvement in the overall project. The municipality of Sölden was the central player, responsible for the construction of the road and drainage infrastructure, as well as making all necessary spatial planning decisions to facilitate development. The state administration supervised these decisions and oversaw the land readjustment procedure.

4 Institutions of Austrian Land Policy

There are various approaches to land policy in Austria. The following section analyses the institutional aspects of land policy manifested in the Kaisers case: (i) active land policy, which involves strategic land acquisition by the municipality; (ii) passive land policy, which is represented by the regulatory instrument of land readjustment, demonstrating how land reallocation is able to promote development; and (iii), procedural land policy, highlighting the role of local land use and development planning in establishing the rights for development.

4.1 Active Land Policy

The term active land policy might not be internationally established with a homogenous understanding but would describe a common practice in Austria. Municipalities strategically buy land to implement their long-term development strategies and to have parcels to bargain with private landowners. For municipalities, this is only possible if they have sufficient financial resources. If municipalities are not in such a fortunate situation the aforementioned public land funds can step in. It is important to highlight that both the municipality and public land funds have to buy land on the free market and do not obtain any preferential treatment by law.

In the Kaisers case the municipality was able to acquire land incrementally and without the support of the state or the land fund. This was possible due to stable political conditions, the broad support from all parties in the municipal council, and the high tax revenues from tourism.

4.2 Passive Land Policy via Land Readjustment

Land readjustment procedures originate from agricultural land consolidation schemes. In the second half of the twentieth century, land consolidation was primarily applied to improve agricultural productivity. Over time, the idea of pooling land and re-distributing parcels while assigning sufficient land for roads and other technical infrastructure was introduced in several Austrian states (BMLF, 1983). They all follow the same procedural layout. Property owners in a clearly demarcated area must confirm their participation in a land readjustment scheme. Depending on the legal prerequisites, the parcels involved may already be zoned for building purposes or may need to be included in the planned expansion perimeter of the urban area according to local planning strategies. Parcels dedicated to agricultural use cannot be included in these schemes. Once the official procedure for readjustment begins, individual property rights are suspended, and all land is pooled.

Subsequently, a land readjustment plan is drafted and negotiated among the participating landowners. A state authority typically oversees and manages the entire process. The participating parties may either collectively agree on the land readjustment plan and waive all rights of appeal, or the readjustment can be approved by a majority vote. If the latter occurs, participants retain their right to appeal, which can potentially lead to lengthy legal disputes.

The advantages for the beneficiaries of readjustment procedures are obvious: (i) instead of unfavourably shaped parcels they receive parcels suitable for development, (ii) technical infrastructure can be provided efficiently with all parties contributing equally, (iii) the cost for the procedure is covered under administrative budgets, leaving only surveying costs and (iv) the increase of property value stays fully with the beneficiaries. Land readjustments are rather typical for the western states of Austria such as Vorarlberg and Tyrol (ÖROK, 2014). The city of Vienna incorporated the instrument in its planning law but has executed only a few procedures. Developers viewed it as an inexpensive way to rearrange plot structures, as the authority would cover most of the costs and manage the process.

In the Kaisers case, the approach was an informal enhancement of the legally established readjustment scheme. The municipality leveraged the existing scheme to secure a key position in negotiations with the other landowners, requesting additional land to be designated for affordable housing.

4.3   Procedural Land Policy

In addition to strategic land acquisition and readjustment, municipalities are responsible for implementing normative spatial planning instruments to regulate land use and guide actual development. As municipalities hold a planning monopoly for local urban development, they need to provide a strategic and long-term vision. Most states require a municipal development concept that defines urban development hot spots, locations for essential facilities, and settlement boundaries. Such strategic considerations are essential for zoning and development plans. It is possible at this stage to limit the use of certain locations for affordable housing, which has to meet strict criteria. At the same time, it is necessary to provide sound and comprehensive justification for such a decision, otherwise such a restriction could be classified as a factual expropriation due to the interference with property rights (Auer, 2011; Lienbacher, 2016). In most Austrian states, municipalities have the authority to zone plots specifically for affordable housing projects, which often involves extensively limiting usage rights. Through subordinate detailed development plans, municipalities can also determine the specific aspects of the design, such as building height and density.

Legally binding planning instruments are typically accompanied by extensive communication efforts, often carried out by the municipalities themselves. To ensure the quality of housing project designs, architectural competitions can be organised. Municipalities often retain the right to nominate renters for the newly built flats, ensuring that those in need have priority access.

In the Kaisers case, the municipality of Sölden identified the location in Kaisers as a potential area for urban development within its development concept, with the stipulation that a land readjustment procedure would be necessary and that affordable housing must be established (Gemeinde Sölden, 2012, 2013). Before enacting the zoning and detailed development plan, an architectural competition was held to select a non-profit housing association willing to undertake the actual development. The municipality retained the right to nominate renters, ensuring that those in need would have the opportunity to move in first.

The Kaisers case illustrates an effective and long-term approach to procedural land policy. It demonstrates the successful combination of active and passive land policy measures, established planning instruments, and informal processes.

5   Reflection

Overall, the Kaisers case provides valuable insights within the context of Austria's planning and property framework:

- Rural municipalities may face challenges typically associated with urban areas, such as significant land use conflicts, high land prices, or strongly limited land available for development. This is especially the case in touristic Alpine municipalities with national and international investments in accommodation and secondary homes.
- Affordable housing for non-landowners must be realised in "B-locations" as central locations are hardly accessible due to property structures and land prices. This entails extensive public investments in infrastructure and—if applicable—in natural hazard protection infrastructure.
- The planning process in the Kaisers case was successful due to the local political environment, which maintained a long-term perspective and stable majorities, allowing for a strategic approach to land policy.
- Lessons learned from previous projects were applied to minimise the financial burden for the municipality, such as opting for a housing association instead of municipal housing, and to ensure the architectural quality through open competitions.

The Kaisers case is ambiguous when it comes to the question of whether it is representative of land policy strategies in Austria. On the one hand, the Kaisers case is exceptional in terms of its unique topography and the resulting limitations on land availability. The Alpine characteristics limit the proportion of the municipal territory suitable for development. The local economy's focus on winter tourism further distinguishes Kaisers. The development of skiing tourism has gradually transformed the former farming village into a small-scale urbanised tourism resort, with urbanisation processes occurring selectively and distant from existing urban areas. Few Austrian municipalities share similar spatial and economic frameworks.

On the other hand, the Kaisers case is representative in terms of institutional aspects of land policy and how they are manifested:

(i) First, there is a strong constitutional protection of property rights in Austria. Strong property rights preserve the existing ownership structures and empower landowners. This is particularly true in Sölden municipality. Given the lack of land supply and the absence of a functioning land market, the municipality is dependent on the willingness of a few landowners to sell their land for public purposes or participate in respective trade-offs. This makes landowners highly influential players in any kind of public land acquisition processes.

(ii) Second, the options of public authorities to interfere with private property rights are limited. It is a feature of the Austrian planning system that zoning—such as designating an area for affordable housing—does not automatically result in implementation, meaning the land is not necessarily made available for the public purpose intended by the zoning regulation. Thus, implementation requires additional land policy measures, which in the Kaisers case meant supporting land supply for affordable housing and providing structural protection against natural hazards. However, measures like pre-emptive rights or compulsory purchase are not widely available to municipalities. These options would be available if existing regulations, like the Land Acquisition Act—a federal law enacted in 1974 providing far-reaching coercive rights in order to facilitate land acquisition for housing—were applied.

(iii) This leads to a third institutional aspect representative of land policy in Austria: the reluctance of political stakeholders to utilise existing regulatory instruments. For land acquisition, the municipality of Sölden focused on negotiations. The land readjustment scheme was based on a consensual agreement of the participating landowners and not on a formally enforced decision by a majority. Due to its long-term political strategy, the municipality was successful. However, this cannot be automatically assumed in other municipalities. The reluctance regarding regulatory instruments also applies to land readjustment schemes as an instrument which, although widely available in Austria, is not often used.

(iv) Fourth, municipal land ownership and land acquisition are highly significant for realising public interests, especially for affordable housing. Providing land for estimated housing demands lies primarily with municipalities. Housing is therefore almost entirely dependent on municipal councils, with minimal involvement or authority from the state or the federal levels (Schindelegger et al., 2023). This turns the municipalities into key players in providing land for affordable housing. Municipal land policy, however, is frequently supported by federal and state authorities. The state government plays an important role as the responsible land readjustment authority. In other cases, state-based land funds may be included if municipalities cannot afford land acquisition. State land transfer regulation, however, can be an obstacle to municipal land acquisition unless the laws provide exceptions for municipalities and land funds to acquire farmland not yet zoned for potential development.

(v) Fifth, with regulatory approaches and financial incentives widely absent, municipalities rely on persuasive instruments to make private land available for public interests. Although land readjustment is a regulatory instrument, it only has a supporting effect and is not able to provide land on its own. Thus, municipal stakeholders work to convince landowners to sell their land. In Austria this approach of strategically acquiring land and buildings for public purposes is referred to as active land policy. Besides affordable housing it is also applied in densification and providing space for green infrastructure. Due to the variety of tasks and different spatial and social framework conditions there is no "one size fits all" approach in active land policy. Municipal stakeholders must develop and pursue individual, tailor-made land policy strategies in order to achieve public interests. First and foremost, however, active land policy requires the willingness of municipal stakeholders to engage in persuasion and lengthy negotiations, tasks which lie outside their manifold administrative obligations.

(vi) Sixth, municipalities committed to active land policy simultaneously act in two roles. On the one hand they are planning authorities and therefore trustees of the common good (Löhr, 2013). On the other hand, they are property owners and beneficiaries of financial revenues attached to development. This constellation may raise concerns regarding the misuse of governmental powers. However, with the exception of the big cities, Austrian municipalities typically do not develop land by themselves; instead, they sell or lease the land to non-profit housing developers. Furthermore, there is no municipal monopoly on selling building land (Hartmann & Spit, 2015); rather, a variety of other suppliers exist, with municipalities generally focusing on public interest development.

Coming back to the understanding of land policy presented in the introduction, the Kaisers case exemplifies the core of Austrian land policy strategies at a local level: developing a tailor-made combination of instruments, including strategic land acquisition to establish affordable housing. This involves crucial negotiations and agreements with landowners. Given the strong position of private landowners in Austria, achieving public interests at a local level may be challenging, but as the Kaisers case demonstrates, it is possible.

References

Alpine Convention. (2016). *The alps eight countries, one territory* (2nd ed.). Permanent Secretariat of the Alpine Convention.

Auer, M. (2011). Zur Entschädigung bei Rückwidmung – OGH versus VfGH: Bemerkungen zu OGH 14.12.2010, 3 Ob 234/10y, OGH 9.9.2008, 5 Ob 30/08k sowie VfGH 4.3.2011 G 13/10. *Baurechtliche Blätter, 14*(4), 168–173. https://doi.org/10.1007/s00738-011-0022-4

Berka, W. (2021). Verfassungsrecht: Grundzüge des österreichischen Verfassungsrechts für das juristische Studium (8th ed.). Verlag Österreich.

Binder, B., Fröhler, L., Lackinger, O., Nowotny, E., Pöll, G., & Zeitlhofer, H. (1990). Bodenordnung in Österreich. Schriftenreihe Kommunale Forschung in Österreich Band 85. Institut für Kommunalwissenschaften und Umweltschutz. Verlag für Jugend & Volk.

BMLF (1983). 100 Jahre agrarische Operationen in Österreich: 1883–1983. Der Förderungsdienst: Sonderheft 1/1983. Bundesministerium für Land- und Forstwirtschaft.

Bundesministerium für Arbeit und Wirtschaft (2023). Tourismus in Österreich 2022. Retrieved September 5, 2024, from https://www.ris.bka.gv.at/Dokumente/Mrp/MRP_20230705_66/009_001.pdf

Gemeinde Sölden (2012). 14. Änderung des örtlichen Raumordnungskonzeptes genehmigt am 16.07.2002 gemäß § 31, TROG 1997.

Gemeinde Sölden (2013). 16. Änderung des örtlichen Raumordnungskonzeptes genehmigt am 16.07.2002 gemäß § 31, TROG 1997.

Hartmann, T., & Spit, T. (2015). Dilemmas of involvement in land management—Comparing an active (Dutch) and a passive (German) approach. *Land Use Policy, 42*, 729–737. https://doi.org/10.1016/j.landusepol.2014.10.004

Kanonier, A., & Schindelegger, A. (2018). Distribution of Areas of Competence and Planning Levels. In ÖROK-Österreichische Raumorndungskonferenz (Eds.), Spatial Planning in Austria with References to Spatial Development and Regional Policy. ÖROK.

Kanonier, A. (2020). Wirkungsfähigkeit von baulandmobilisierenden Instrumenten im Raumordnungsrecht. *Baurechtliche Blätter, 23*(4), 119–135.

Kraus, S. (2018). Kommunale Bauland- und Bodenpolitik. Dargestellt anhand von Bodenfonds, der Baulandumlegung, städtebaulichen Verträgen und der Planwertabgabe. Masterarbeit am Institut für Raumplanung, Umweltplanung und Bodenordnung, Universität für Bodenkultur Wien.

Kühne, J. (Ed.) (1982). Eisenbahnenteignungsgesetz: unter besonderer Berücksichtigung allgemeiner Enteignungsgrundsätze; samt einschlägigen Rechtsvorschriften sowie mit Einführungen, Erläuterungen, Anmerkungen und Rechtsprechung. Manz.

Lienbacher, G. (2016). Raumordnungsrecht. In Bachmann, S., Baumgartner, G., Feik, R., Fuchs, C., Giese, K. J., Jahnel, D., & Lienbacher, G. (Eds.), Besonderes Verwaltungsrecht: Lehrbuch (11th ed.). Verlag Österreich.

Löhr, D. (2013). Vom Aschenputtel zur attraktiven Braut: Das kommunale Erbbaurecht. In Senft, G. (Ed.), Land und Freiheit. Zum Diskurs über das Eigentum von Grund und Boden in der Moderne, 190–198. ProMedia.

Mayer, K., Ritter, K., Fitz, A., & Architekturzentrum Wien (Ed.) (2020). Boden für alle. Park Books.

Mayer, K. (2020). Die Ware Boden…oder warum Boden kein Joghurt ist. In Mayer et al. (Eds.), Boden für alle (pp. 60–83). Park Books.

ÖROK - Österreichische Raumordnungskonferenz (Ed.) (1993). Wirksamkeit von Instrumenten zur Steuerung der Siedlungsentwicklung. Schriftenreihe Österreichische Raumordnungskonferenz Nr. 105, Wien.

ÖROK - Österreichische Raumordnungskonferenz (Ed.) (1995). Möglichkeiten und Grenzen integrierter Bodenpolitik in Österreich. Schriftenreihe Österreichische Raumordnungskonferenz Nr. 123, Wien.

ÖROK-Atlas (2015). Anteil des Dauersiedlungsraums in den Gemeinden an der Gesamtfläche 2015. Retrieved July 16, 2024, from https://www.oerok-atlas.at/#indicator/74

ÖROK - Österreichische Raumordnungskonferenz (Ed.) (2014). Beiträge der Raumordnung zur Unterstützung „leistbaren Wohnens": Ergebnisse der ÖREK-Partnerschaft. Schriftenreihe Österreichische Raumordnungskonferenz Nr. 191, Wien.

Riedmüller, M., & Erhart, A. (2020). Der Tiroler Bodenfonds als Instrument der aktiven Raumordnung. *Baurechtliche Blätter, 23*(5), 184–186. https://doi.org/10.33196/bbl202005018401

Schindelegger, A., Seher, W., & Gutheil-Knopp-Kirchwald, G. (2023). How to mobilise land for affordable housing in transforming urban areas: approaches linking land policy with housing policy in Austria. In ENHR European Network of Housing Research (Ed.), *Urban regeneration - shines and shadows, book of Abstracts*.

van Straalen, F.M. (2014). Private property in public processes: How public stakeholders strategically interfere in private property rights in the public interest in regional spatial development processes [Dissertation. Wageningen University].

Arthur Schindelegger is a postdoc research fellow at BOKU University, Institute of Landscape Planning. He has studied spatial planning at TU Wien and KTH Stockholm and has worked in the private sector for several years. His research is centred around questions on land policy, planning law and the integration of natural hazards and climate change adaptation in planning processes and procedures. He is also a consultant with the World Bank on resilient urban development.

Walter Seher is assistant professor and head at the Institute of Spatial Planning, Environmental Planning and Land Rearrangement at BOKU University. He holds a master's degree in civil engineering and water management and a PhD from BOKU University, Vienna. His key qualifications include spatial planning in natural hazard risk management and climate change adaptation, land policy and land rearrangement. He is general secretary of the European Academy of Land Use and Development (EALD).

Land Policy in Belgium: How to Limit Land Take in a "Landowners' Paradise"?

Jean-Marie Halleux and Hans Leinfelder

Abstract In Belgium, the relationship between individual property rights and collectively desired land use objectives has historically favoured landowners, leading to the country being described as a "landowners' paradise". However, this advantageous position for landowners presents a serious obstacle to controlling urban sprawl. In response to the European Union's goal of limiting land take, Belgian authorities are now working to develop new policies that align with the no-net-land-take (NNLT) objective. This applies to the two main regions of the country: Flanders and Wallonia. Through the exploration of two cases in the Flemish city of Ghent, this contribution assesses how Flanders' planning is designing strategies to meet the NNLT goals. A reflection on the similarities and differences with the policy approach in Wallonia to meet almost the same goals holds international relevance, as it demonstrates how two regions emerging from a common legal and planning framework can adopt substantially different strategies due to their respective contextual factors.

1 Introduction: The Belgian Understanding of "Land Policy"

Two main characteristics are prerequisites for understanding how land policy and spatial planning operate in Belgium. The first relates to a cultural and political context in which the planning tradition is notably weak, especially when compared to neighbouring countries such as the Netherlands, Germany, and France. In Belgium, the historical relationship between individual property rights and collectively desired land uses has consistently favoured landowners (Halleux et al., 2012). As a result,

J.-M. Halleux (✉)
University of Liege, Liege, Belgium
e-mail: jean-marie.halleux@uliege.be

H. Leinfelder
KU Leuven, Ghent, Belgium
e-mail: hans.leinfelder@kuleuven.be

© The Author(s) 2025   35
T. Hartmann et al. (eds.), *Land Policies in Europe*,
https://doi.org/10.1007/978-3-031-83725-8_3

while the Netherlands is often characterised as a "planners' paradise", Belgium can be described as "landowners' paradise" (Shahab et al., 2021).

This advantageous position for landowners comes at a significant cost, particularly in terms of controlling urban sprawl. For instance, when comparing artificial surfaces to population or GDP size, Belgium's land consumption is markedly above the European average (Bengs and Schmidt-Thomé, 2006). The high level of urban sprawl in Belgium poses several challenges, including landscape fragmentation, increased car dependency, and rising public infrastructure costs (see European Environment Agency and Federal Office for the Environment (2016) for an extensive comparison of urban sprawl in European countries).

The second characteristic of the Belgian context is its status as a federal state, where spatial planning, including land policy, is the exclusive responsibility of four federated entities: the Brussels-Capital Region, the Flemish Region, the Walloon Region, and the German-speaking Community. This division means that four distinct planning systems coexist within Belgium (Halleux & Lacoere, 2023b), despite their origins in a common framework established by the national government in the 1960s. It was only in the 1980s that the planning systems and legislation of the federated entities began to diverge, reflecting the transition from a unitary to a federal state.

In this contribution, we will focus on the Dutch-speaking region of Flanders and the French-speaking region Wallonia. This focus is justified not only by the fact that those two entities together represent 97% of the national territory, but also because they are the most actively engaged in the ambition of no-net-land-take (NNLT), which is the central theme of this discussion. Recent strategic planning documents from both regions have articulated a strong political commitment to significantly reduce additional land take in pursuit of reaching land take neutrality. However, the necessary coherent legal framework for operationalising this NNLT is still lacking in both Flanders and Wallonia.

Our contribution is based on two case studies conducted in the Flemish city of Ghent. By examining these cases, our first objective is to assess how the planning system in Flanders designs solutions to achieve the NNLT ambition. Simultaneously, our second objective is to explore the similarities and differences in policy approaches between Wallonia and Flanders in addressing nearly the same NNLT ambition. This comparative analysis is particularly relevant in an international context, as it highlights how approaches within similar land policy systems—emerging from a shared Belgian planning matrix—can differ significantly due to political, financial, and institutional factors. This underscores the critical importance of thoroughly evaluating these contextual factors before attempting to replicate land policies from one country to another.

From the outset, it must be acknowledged that land policy in Belgium is often seen as a "forgotten necessity" (Doucet, 1985). Both Flemish and French-speaking analysts have long criticised this deficit as a fundamental weakness in the functioning of the Belgian planning systems (Doucet, 1985; Hendrickx, 2001).

To understand the term "land policy" in Belgium, it is helpful to examine two key definitions: the Flemish concept of *grondbeleid* (in Dutch) and the Walloon concept of *politique foncière* (in French). The first definition appears in Flanders' initial regional strategic spatial planning document (*Ruimtelijk Structuurplan Vlaanderen*; Vlaamse overheid, 1997; translation by authors): "Land policy is the set of measures related to the acquisition, provision, and management of land parcels in order to promote planning policy objectives. An appropriate land policy (…) avoids, for example, speculation on these plots." The second definition, from French-speaking scholar and lawyer Francis Haumont (1990; translation by authors) states: "Land policy is the set of actions implemented in a combined way, by which the public authorities endeavour to dispose of land by influencing, in varying degrees, the behaviour of its owners and users—if necessary by counteracting them, opposing them, capturing at least partially the increase in land values or even by making themselves land owner while being ready, when it is appropriate, to compensate the decline in land value."

These definitions highlight three core dimensions that characterise an ideal land policy in Belgium. First, there is the need for a cohesive policy that integrates the objectives of various sectoral policies. Second, the territorial dimension is crucial, as both definitions emphasise the importance of spatial planning. Finally, there is an economic and financial aspect, addressing issues like land speculation, value capture, and compensation for changes in land values.

Despite these conceptual definitions, there remains a significant gap between the theoretical understanding of land policy and its practical implementation. Neither Flanders nor Wallonia currently have a consistent, overarching approach to land policy. This is not to say that Belgian authorities are inactive in managing land; rather, they use various land management instruments, such as expropriation, emphyteusis right, and planning obligations, which are integrated in the Civil Code as well as in sector-specific legislation (e.g., environment, agriculture, and industrial development). The use of these instruments varies depending on the policy domain. For instance, public authorities have been highly active in the development of industrial land since the 1960s, often through inter-municipal companies (Vandermeer & Halleux, 2017). In contrast, public intervention in the development of residential land has historically been limited in Belgium (Heynen, 2010). As a result, most residential development is initiated by private landowners and/or developers. This limited public intervention in the housing markets is largely due to strong private property rights and to an abundance of potential building land outlined in the sub-regional land use plans (*plans de secteur* in French and *gewestplannen* in Dutch) (see below in Sect 2.2).

2   Belgian Land Policy in Practice: Akkerstraat and Slotendries in Ghent

The city of Ghent is part of the urban network known as the Flemish Diamond (Albrechts & Lievois, 2004). With nearly 270,000 residents, the city anchors Belgium's fourth most populated agglomeration. Ghent has a historical, densely built city centre, encircled by nineteenth century industrial neighbourhoods. Additionally, it features an urban fringe composed of agricultural zones, natural areas connected to the Lys and Scheldt rivers, and scattered urban sprawl characterised by both residential and economic development. This makes Ghent a prime example of the typical Flemish "urban nebula" (*"nevelstad"*; Dehaene & Loopmans, 2003) or "horizontal metropolis" (Vigano et al., 2018), terms that describe the dispersed, low-density nature of urban sprawl in Flanders.

2.1 *"Land Use Allocation Neutrality" at the Municipal Level*

In 2018, Ghent's city council adopted a new strategic spatial structure plan titled "Space for Ghent" (*Ruimte voor Gent*; Stad Gent, 2018), a long-term planning vision emphasising the importance of green and blue networks within and around the city. As a part of this plan, the city council introduced the concept of *ruimteneutraliteit,* or "spatial neutrality". In practical terms, this means immediately halting any further "hard" land use allocations—such as those for residential, industrial, recreational, or infrastructural purposes—within the municipality. The goal is to prevent these developments from encroaching on land currently allocated for agricultural or nature preservation. A more accurate translation of *ruimteneutraliteit* might be "land use allocation neutrality".

According to this principle, any new land required for development must be found within existing urban allocations. If, in exceptional cases, it is deemed necessary to convert agricultural, natural, or forested land into residential or industrial land, this shift must be compensated by an equal amount of land being reallocated in the opposite direction—within the city's own territory. This planning principle aligns with the broader aim of achieving NNLT. However, it is important to note that "land use allocation neutrality" is only a necessary, though not sufficient, condition to fully meet NNLT goals, as will be discussed further in this chapter.

In 2021, the city council took another significant step by approving a thematic municipal land use plan "169 Green" (*gemeentelijk ruimtelijk uitvoeringsplan 169 Groen*). The purpose of this plan was to reinforce the city's green infrastructure by re-zoning 101 small areas located across Ghent, resulting in an additional 370 ha of land designated for nature and forestry—an increase of 20% in the city's green spaces. To illustrate the impact of this policy, we will focus on two representative areas that were re-zoned under the new plan: Akkerstraat and Slotendries (see Fig. 1).

Akkerstraat is a small, ecologically valuable wooded area spanning 0.36 ha in the city centre (see Fig. 2). It is one of few unsealed plots within a densely populated residential neighbourhood. Originally designed for residential development in the 1970s under a sub-regional land use plan, the plot has been subject to multiple subdivision applications by its owner. Despite existing Flemish legislation that protect wooded areas, these applications were based on the old land use designation. To prevent further attempts to develop this land, the *169 Green* plan reclassified it as forestry area, ensuring its protection.

Fig. 1 Akkerstraat and Slotendries in Ghent, Belgium (map: Kleiner, Jehling 2024, Aerial imagery provider © Esri, Beeldmateriaal.nl, Maxar, Microsoft)

Slotendries is a much larger area, covering nearly 20 ha on Ghent's eastern urban fringe (see Fig. 3). It includes a forest linked to a castle park, a significant portion of the park itself, and land surrounding a pilgrimage site. There is a political will to extend it towards and beyond the city's major ring road. The castle's moat and park are considered ecologically significant, and the open space serves as a crucial green corridor, linking the city centre with a future recreational green space. In the 1970s, the sub-regional land use plan allocated Slotendries primarily for community facilities—likely in connection with a nearby school campus south of the castle park—and for residential development. However, the *169 Green* plan has now re-zoned the area for forestry and park use, preserving its ecological and recreational functions.

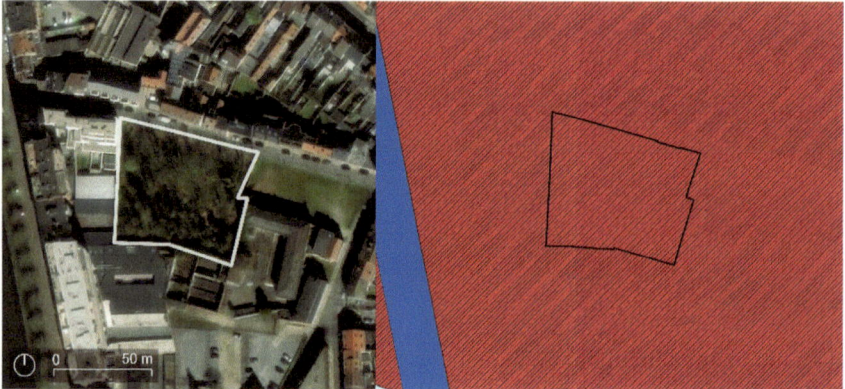

Fig. 2 Perimeter of Akkerstraat small wood (map: Kleiner, Jehling 2024, Aerial imagery provider © Esri, Beeldmateriaal.nl, Maxar, Microsoft) and initial designation of Akkerstraat as a residential area (coloured red) in the sub-regional land use plan

Fig. 3 Perimeter of Slotendries (map: Kleiner, Jehling 2024, Aerial imagery provider © Esri, Beeldmateriaal.nl, Maxar, Microsoft) and allocation for community facilities (coloured in light blue) and residential development (coloured in red) in the initial sub-regional land use plan

2.2   A Cultural Context of Deep Respect for Property Rights and a History of Abundant Land Supply in Land Use Plans

Belgium's weak tradition of spatial planning in general has deep historical roots and is strongly influenced by the country's emphasis on private property rights and the resulting limitations on active land policies (see Lacoere & Leinfelder, 2022). This situation also stems from the fact that it wasn't until 1962 that the national government enacted formal planning legislation. The 1962 Planning Act outlined a framework for creating land use plans on four levels: regional, sub-regional, municipal, and sub-municipal. While no regional plans were ever produced, a clear impetus was given in 1964 when the national government began preparing sub-regional land use plans to cover the entire country. This move was prompted by a lack of enthusiasm for planning at the municipal level.

In 1962, the sub-regional plans were envisaged merely as a coordination document between the regional plan and the municipal plan. Despite this initial conception, the sub-regional land use plans, which were approved in the 1970s and 1980s, have in fact become and are still the most influential legally binding type of land use plans in Flanders and Wallonia. In both regions, any planning permit must comply with these plans unless a lengthy modification process is undertaken. In Wallonia, the sub-regional land use plans still cover the whole region, while in Flanders they provide the framework for planning policy across roughly 80% of the territory (Pisman et al., 2018). The remaining 20% has been rezoned through subsequent regional, provincial, or municipal spatial implementation plans (*ruimtelijke uitvoeringsplannen*), which are similar in structure and purpose to the original sub-regional land use plans. The fact that the federated entities are responsible for drafting these binding land use plans appears to be quite specific to the Belgian national context. Indeed, in most neighbouring countries, such plans are drawn up by local authorities, not by national or regional authorities (ESPON, 2018).

One notable characteristic of these sub-regional land use plans is the significant oversizing of residential zones between the 1960s and 1980s. This was likely due to two factors; first, the plans were drawn up during a time of anticipated economic and demographic growth; second, the 1962 Planning Act introduced a compensation mechanism for downzoning. Fearing large compensation payouts, planners were "generous" in designating large greenfield areas for residential use to avoid legal challenges by landowners who might claim that their land had been considered for development prior to the plan's implementation (Haumont, 1990; Lacoere & Leinfelder, 2020).

The oversizing of residential zones is one of the primary causes of the extensive urban sprawl that characterises much of Belgium today. Despite changing urban dynamics, greenfield residential zones remain overabundant in most parts of the country. The introduction of the compensation principle in 1962, driven by property interests and a cultural respect for property rights, contributed significantly to this pattern. Historically, Belgium's approach to balancing individual property rights with collective land use goals has favoured landowners (Halleux et al., 2012).

2.3   The Current Political Objective to Stop Additional Land Take

The Flemish regional government has adopted the goal of achieving NNLT as part of its strategic vision for a new Spatial Policy Plan for Flanders (*Strategische Visie*

Beleidsplan Ruimte Vlaanderen; Vlaamse overheid, 2018). In line with the European Union's NNLT target (European Commission, 2016), the original target date was set for 2050. However, following public criticism, the Flemish government moved the deadline forward to 2040. This policy, coined by the media as "concrete stop" (*betonstop*), was later officially renamed as "construction shift" by the government (*bouwshift*).

In 2019, the Walloon government adopted its own version of the "concrete stop" (*stop béton*) as part of its Territorial Development Scheme (*SDT* or *Schéma de développement du territoire*; Gouvernement Wallon, 2019). This initiative was partly in response to both the European agenda and the Flemish government's goals. The 2019 SDT aimed to cut land consumption by half, from 12 to 6 km^2/year by 2030, and moving towards net zero land take by 2050. However, the Walloon government did not set a firm date for when these measures would take effect, and the document was finally withdrawn in 2022.

In April 2024, the Walloon Government approved a revised version of the SDT, with the measures officially to enter into force in August 2024. The NNLT objective remains similar to the one in the 2019 plan, although the interim milestone for 2030 is no longer included. The more neutral term "spatial optimisation" (*optimisation spatiale*) has been adopted to describe the policy. In addition to the NNLT target, the Walloon authorities also aim to curb the spatial dispersion of new residential developments. By 2050, at least 75% of new dwellings are expected to be located within designated "urban and rural centralities". This is intended to prevent further expansion of remote residential developments. However, ongoing discussions are still taking place regarding the exact delimitation of the "centralities" and the different regulations that will apply to planning applications inside and outside these areas.

2.4 Important Differences Between Types of Neutralities

The City of Ghent's adoption of "spatial neutrality" may seem straightforward, but this concept can be misleading. The term suggests that no additional agricultural or natural land would be developed for residential or economic purposes. However, upon closer examination, "spatial neutrality" actually refers to *land use allocation neutrality*. This means that any new allocations for residential, economic, recreational, or infrastructural development are offset by reallocations of land within the same administrative area for nature, forestry, or agriculture.

In 2022, Ghent's pioneering concept of "land use allocation neutrality" was incorporated into regional legislation by the Flemish government in the form of mandatory "planning compensation" (*planologische compensatie*). It is important to distinguish this form of "planning compensation" from the more common international interpretation, where the term refers to financial compensation for landowners in cases of downzoning. To avoid confusion, we will continue to use the more specific term "land use allocation neutrality".

Despite its innovative character, land use allocation neutrality alone does not prevent further land take. This is largely due to the significant amount of land designated for residential development in sub-regional land use plans. Much of this designated land is still undeveloped and, in reality, continues to be used as farmland. Downzoning it in land use plans to agricultural area, to offset the conversion of agricultural area into residential area elsewhere, is technically indeed "land use allocation neutral", but does not achieve a situation of NNLT. This is why the Advisory Board on Planning Issues in Ghent notes that simply implementing land use allocation neutrality is insufficient to reach the NNLT goal (Gecoro, 2020).

NNLT demands more than just balancing land use allocations. It requires that all new residential, economic, recreational, and infrastructural development occur entirely within existing settlement areas. Achieving NNLT involves strategies such as intensification, densification, mixed land uses, and the redevelopment of abandoned industrial sites. If these strategies prove impossible, any additional land use for new development must be offset by the conversion of existing developed areas into open space. This approach results in zero or no net land take.

Furthermore, the Ghent advisory board encouraged the city council to go even beyond NNLT, and pursue a *net gain of open space*. This more ambitious goal would involve identifying poorly located or scattered buildings and demolish them to restore open space, regardless of how those areas are designated in land use plans. The strategy prioritises a qualitative approach to open space in Ghent, aiming to actively reclaim land and enhance the city's natural and agricultural landscape.

3 Relevant Actors for NNLT in Akkerstraat and Slotendries

The main private and public actors relevant for Akkerstraat and Slotendries include private landowners, government authorities, and nature conservation organisations.

3.1 Private Actors: A Landowners' Paradise

Akkerstraat and Slotendries areas are predominantly owned by small, private landowners, each holding one or a few plots, often scattered across the region. This fragmented ownership is a common feature of the Belgian real estate market, which is dominated by numerous private landowners rather than large construction firms, unlike neighbouring countries such as the Netherlands. Notably, only a few plots in Slotendries are owned by the City of Ghent, reflecting another typical aspect of the Belgian context—public authorities own relatively little land.

The allocation of land use to every square meter of the Belgian territory during the 1970s and 1980s created a significant divergence in real estate value, particularly for farmland. Land designated for residential or economic activities under sub-regional land use plans saw a sudden increase in market value, without any governmental taxation on this added value. Conversely, a change in zoning that reduced land value, such as zoning from residential or commercial to agricultural use, is politically and socially unpopular and rarely pursued. Recently, this resistance was reinforced by changes to Flemish regulations, which now provide an even higher substantial financial compensation to landowners for losses in real estate value due to downzoning.

3.2   Public Actors and Land Policy

In Flanders, land use planning operates across three levels of government: the regional (Flemish), provincial, and municipal. The authority to draft a spatial implementation plan—essentially a land use plan—is determined through negotiations between these levels, following the principle of subsidiarity. Subsidiarity implies that decisions should be made at the most local level possible unless the complexity or scope of the issue demands higher-level intervention. For example, the Flemish government would draft a land use plan for the expansion of a harbour, a province for a recreation park, and a municipality for additional housing (Halleux & Lacoere, 2023).

In the case of Akkerstraat and Slotendries, a municipal land use plan was needed to safeguard and develop local nature features. This plan had to align with the city's broader spatial strategy, "Space for Ghent", as well as provincial and regional frameworks. While the provincial and Flemish governments provided advisory input, their approval was not required. The final decision rested with the City of Ghent, which was expected to consider the advice from higher-level authorities, but retained autonomy over the plan.

The City of Ghent's municipal planning department either prepares land use plans or commissions them to private planning firms. The planning department also coordinates input from various municipal departments, such as the departments of public greenery and agriculture. For more complex projects, other departments may be involved as well. Like an increasing number of larger municipalities, Ghent operates its own urban development agency (*stadsontwikkelingsbedrijf*), which has a high degree of independence and flexibility. This agency participates actively in the real estate market, buying, selling, and developing land on behalf of the city.

3.3   Non-governmental Organizations

Another important actor in the development of the municipal land use plan is Natuurpunt, a non-governmental nature conservation organisation. Formed from the merger of two older nature conversation organisations dating back to the 1930s and 1950s, Natuurpunt buys and manages land for conservation purposes. The organisation receives substantial subsidies from the Flemish government, which relies on Natuurpunt to help meet its international and regional conversation targets. Many of the plots whose designations were altered by the municipal land use plan had already been purchased by Natuurpunt, often with these government subsidies, and were converted from farmland into nature reserves. In the future, Natuurpunt plans to acquire additional plots affected by the land use plan, as their prices will decrease since land is rezoned from agricultural use to conservation areas.

4   Institutions in Relation to NNLT in Akkerstraat and Slotendries

4.1   Active Land Policy

In Belgium, the conventional approach to facilitating new residential development involves drafting and approving a land use plan, then waiting for developers to take the initiative. This passive approach relies heavily on landowners' willingness to sell their land. However, under certain circumstances, more proactive measures can be adopted by the city and regional government. For example, in Akkerstraat and parts of Slotendries, the city council of Ghent has adopted a pre-emption right perimeter, enabling the city to acquire forested or nature areas when they come up for sale.

Disadvantage of this strategy is that it hinges entirely on the landowner's willingness to sell. To address this limitation, Ghent's city council has also approved expropriation plans for key parts of the areas covered by the municipal land use plan, such as the castle park and other plots formerly zoned for residential development. Expropriation ensures the city can acquire these plots, which is critical for implementing necessary development and management measures. These include making public areas accessible, removing existing constructions, and afforesting certain plots.

It is important to note that Belgian public authorities are generally hesitant to expropriate private land. When they do, it is typically for industrial development, with expropriation for nature development or social housing being quite rare.

4.2   Reactive Land Policy

As mentioned earlier, changes in land use allocation via municipal land use plans directly affect the real estate value of a plot. Under the Belgian Planning Act of 1962, a compensation scheme was introduced for landowners whose properties were affected by planning decisions that imposed building prohibitions. Compensation was set at 80% of the difference between the indexed purchase value and the market value after the land use change, but this only applied to the first 50 m of plots located along equipped roads. To claim this compensation, landowners had to go to court, leaving the financial decision to the judge's interpretation, which created uncertainty. Crucially, this right to compensation was, and remains, without a time limit. This means that even today, several decades after the sub-regional plans came into effect, downzoning the designated building land could still result in compensation.

However, the Flemish government has recently revised how this compensation is calculated. Intense public debate arose over compensations related to the downzoning of residential areas for the protection of existing forests. In response, the new Instruments Decree (*Instrumentendecreet* May 26, 2023) was enacted, which now provides 100% compensation for the difference between the market value before and after the land use change. Additionally, the procedure for claiming this compensation has been simplified, making it easier for landowners to pursue.

Simulations suggest that this new compensation scheme could result in payments of up to an astonishing tenfold increase when compared to the old system. As a result, municipalities are now hesitant to adopt new land use plans that support NNLT policies, as the financial burden could be enormous (Lacoere, 2023). The City of Ghent acted strategically by approving its municipal land use plan "169 Green" before the new decree took effect. Moreover, by carefully delineating the land use zones by focusing on the first 50 m along equipped roads, the city minimised potential compensation claims in areas like Akkerstraat and Slotendries.

The increased compensation scheme poses significant challenges towards the feasibility of the Flemish NNLT policy. The region faces a considerable oversupply of zoned but undeveloped land under sub-regional plans, which do not expire. GIS analyses have shown that at least 30,000 ha of zoned land need to be downzoned, with 11,000 ha requiring urgent protection due to climate and nature conservation goals (Bouckaert et al., 2021). However, with compensation costs rising significantly, future initiatives to meet NNLT targets will be more difficult to implement.

4.3   Passive Land Policy

Land use plans are inherently passive planning instruments. They designate how land can be used but do not actively engage in reshaping land ownership or development dynamics. However, the 2023 Instruments Decree introduced a new mechanism called "planning trade", aimed at making land use planning more active by combining it with land readjustment measures. Officially called "land readjustment by law with a land use plan" (*herverkaveling uit kracht van wet met een ruimtelijk uitvoeringsplan*), this mechanism pairs downzoning operations with proportional upzoning at a more suitable location. At the same time, it allows for government-led land ownership transfers between these areas. Due to its recent introduction however, this planning trade mechanism has not yet been applied in Flanders.

Another recent policy shift is the Decree on Residential Reserve Areas (*Decreet woonreservegebieden*), which was approved by the Flemish Parliament on the same day as the Instruments Decree. This decree essentially places residential reserve areas under a "glass bell", meaning they should remain untouched unless the municipal council decides to "free" them for development. Since this instrument was introduced after the approval of the municipal land use plan, the City of Ghent has not yet evaluated its potential.

5 Reflection on Differences and Similarities Between Flemish and Walloon Approaches in Land Policy

Although both the Flemish and Walloon planning systems are based on the Belgian Planning Act of 1962, they have diverged significantly over the last two decades. Despite sharing the same legal foundation, after 40 years of "living apart together", the two regions have developed different approaches to land use and achieving NNLT goals, shaped by distinct geographical, political, and economic contexts. This final chapter reflects on these differences and similarities, using the Ghent case study and Flanders's approach to land policy as a lens to compare with Wallonia's land policy.

5.1 Land Value Compensation: From a Landowners' Paradise to a Super Landowners' Paradise

The recent adjustments to the planning compensation mechanism in Flanders have significantly increased the costs for governmental budgets, particularly due to compensation claims for downzoning. From the point of view of private landowners, this shift has turned Flanders from a landowner's paradise into a *super* paradise. However, these generous compensation mechanisms contradict proposals to achieve NNLT goals by 2040.

This fundamental change in land value compensation has not taken place in Wallonia. On the one hand, the landowners' right to compensation is not questioned. On the other hand, the economic and political conditions to further support the landowners are not fulfilled. Public budgets in Wallonia are, at all policy levels, less favourable than those in Flanders. That is why, at this moment, it seems unthinkable that political parties in the Walloon government and parliament would modify the existing compensation mechanism. This would put a large burden on the municipal and regional budgets. From this perspective, it is clear that the Flemish discussion about the compensation mechanism is more easily facilitated by the fact that Flanders is one of the most prosperous regions in Europe.

Another explanation for the status quo in Wallonia relates to the political balance. During the last decades, the political coalitions in both regions were quite different: where the majority in Flanders was situated at the right of the political centre, the majority of Wallonia was situated at the left. Defending the rights of individual landowners appeared less in line with the predominant political forces in Wallonia.

While it may be feasible from a strictly legal perspective, such a policy could jeopardise electoral support (Lacoere et al., 2023). As a result, when considering NNLT, planners in Wallonia are driven to seek innovate solutions such as delineating "urban and rural centralities". Nevertheless, as mentioned in Sect. 2.3, there are ongoing debates about the differential of treatment of planning applications inside and outside of these areas. Moreover, the issue of oversupply of legally zoned but undeveloped land outside these centralities remains unresolved.

5.2   Land Use Allocation Neutrality as a First Step Towards NNLT

As discussed in Sect. 2.4, land use allocation neutrality is only a first step in developing a NNLT policy. This neutrality is theoretical on paper but does not immediately translate to neutrality in the field. Achieving true neutrality would require preventing further loss of open space and, where unavoidable, restoring open space elsewhere.

Flanders has recently introduced mandatory "land use allocation neutrality", but Wallonia has already had a similar provision in place since 2005 (*compensation planologique*). The Walloon experience confirms that land use allocation neutrality alone is insufficient to curb land take. Despite this policy, significant land consumption has occurred in Wallonia since 2005, largely due to the sub-regional land use plans that still classify large greenfield areas as buildable land. If there had been less undeveloped but zoned land, strict adherence to land use allocation neutrality might have yielded different results.

5.3   The Two Sides of NNLT Policy

The Ghent case studies illustrate the protective side of NNLT policy, aimed at safeguarding open spaces. The other, more commonly embraced aspect of NNLT involves promoting the development of strategically situated plots within cities, towns, and villages. Both Walloon and Flemish regional and municipal governments are particularly supportive of this second aspect, which focuses on densification, mixed land uses, and the reconversion of abandoned or outdated sites within established urban areas.

However, Belgium's overall surplus of building land in sub-regional land use plans makes the protective side of NNLT particularly challenging. The Ghent case studies, which feature municipal downzoning initiatives, must be regarded as exceptional. Indeed, political and technical complexities, combined with the risk of expensive compensations claims, mean that downzoning initiatives remain rare. Moreover, these efforts have primarily focused on plots suitable for residential use or public use, while significant amounts of greenfield land allocated for economic and industrial activities remain untouched.

At the regional level, efforts to address the oversupply of building land and strengthen NNLT protections are still in their early phases, both in Flanders and Wallonia. In Wallonia, the 2024 SDT encourages municipalities to attract new investments within designated "centralities", but how to operationalise this vision through planning applications remains unclear. Similarly, Flanders has introduced a comparable strategy under its recent Decree on Residential Reserve Areas, which places undeveloped residential zones under a protective "glass bell" (see Sect. 4.3). However, how this will influence future planning applications remains unclear.

5.4 Differences in Strategies and Institutional Power of Municipalities

From a planning perspective, large Belgian cities like Ghent exhibit two major distinctions when compared to smaller municipalities on the urban outskirts or in rural areas with a few thousand of inhabitants. First, these cities generally agree with the regional objectives to limit urban sprawl and to achieve the NNLT ambition. Second, they are better equipped in terms of institutional power—which includes human resources, financial means, and political influence especially in relation to the regional government. In this context, the NNLT strategy developed by the City of Ghent should not be considered as a common municipal strategy.

In contrast, most suburban and rural municipalities are more inclined to promote continued residential and economic development, leading to further urbanisation. This is driven by various factors, chief among them being the associated increases in municipal tax revenues. For this reason, public policy aimed at achieving NNLT must extend beyond traditional planning measures and include fiscal and budgetary policies designed to adjust the development incentives for all types of municipalities. For example, introducing alternative revenue streams from the regional government—particularly for municipalities with substantial agricultural or nature reserves—could reduce their incentive to encourage new land take.

6 Reflection

By examining two specific cases within a single municipality in Flanders and reflecting on the differences and similarities between the Flemish and the Walloon approaches, this chapter illustrates the land use policies that both Belgian regions have developed to meet their own and international NNLT policy ambitions.

Our most striking observation is that, despite their shared origins in a common Belgian legal framework, the approaches to land policy in Flanders and Wallonia have diverged significantly, largely due to differences in institutional, political, and financial contexts. We believe this observation is also relevant on an international level, as it highlights the importance of thoroughly assessing contextual factors before attempting to transfer land policies between countries or even between regions.

Despite these clear differences in land policy approaches, both regions have faced significant challenges in implementing effective NNLT policies. This is particularly evident within the broader context of what we have termed a "landowners' paradise". We have again observed how the combination of an oversupply of land zoned for urban uses, the time-consuming and judicially complex procedures required to modify these land use plans, and the generous compensation schemes for private landowners, reates significant barriers to active land policy reform in relation to NNLT. This also explains why the Flemish and Walloon regional governments, along with provincial and the municipal authorities, continue to pursue a passive approach to land policy that offers little to no prospect for decisive NNLT implementation.

References

Albrechts, L., & Lievois, G. (2004). The flemish diamond: Urban network in the making? *European Planning Studies, 12*(3), 351–370. https://doi.org/10.1080/0965431042000195038

Bouckaert, J., Lacoere, P., Paelinck, M., & Tindemans, H. (2021). Taskforce Bouwshift. Eindadvies. Retrieved from https://omgeving.vlaanderen.be/nl/rapport-van-de-taskforce-bouwshift-beschi kbaar

Dehaene, M., & Loopmans, M. (2003). De argeloze transformatie naar een diffuse stad: Vlaanderen als nevelstad. AGORA Magazine, 19(3), 4–6. https://doi.org/10.21825/agora.v19i3.3690

Doucet, P. (1985). La politique foncière: Une nécessité oubliée ? *Les Cahiers De L'urbanisme, 6*, 65–78.

ESPON (European Observation Network for Territorial Development and Cohesion) (2018). *COMPASS—Comparative Analysis of Territorial Governance and Spatial Planning Systems in Europe—Final Report.* ESPON EGTC.

European Commission (2016). *No Net Land Take by 2050?* Retrieved October 11, from https://cat alogue.unccd.int/650_no_net_land_take_by_2050.pdf

European Environment Agency and Federal Office for the Environment (2016). *Urban Sprawl in Europe.* Joint EEA-FOEN report. Retrieved October 11, from https://www.eea.europa.eu/pub lications/urban-sprawl-in-europe

Gecoro (2020). Advies bestemmingsneutraliteit. Retrieved October 11, from https://stad.gent/nl/ over-gent-stadsbestuur/stadsbestuur/speel-een-rol-het-beleid/ik-wil-meedenken/adviesraden/ gemeentelijke-commissie-voor-ruimtelijke-ordening#Gecoro-adviezen

Gouvernement Wallon (2019). Schéma de Développement du Territoire. Retrieved from https:// lampspw.wallonie.be/dgo4/tinymvc/apps/amenagement/views/documents/amenagement/reg ional/sdt-v2/1-sdt/sdt-definitif-adopte-16-mai-2019-fr

Halleux, J. M., & Lacoere, P. (2023). *Belgium.* Retrieved October 11, from https://www.arl-intern ational.com/knowledge/country-profiles/belgium

Halleux, J. M., Marcinczak, S., & van der Krabben, E. (2012). The adaptive efficiency of land use planning measured by the control of urban sprawl. *The cases of the Netherlands, Belgium and Poland. Land Use Policy, 29*(4), 887–898. https://doi.org/10.1016/j.landusepol.2012.01.008

Haumont, F. (1990). Les instruments juridiques de la politique foncière. Story-Scientia

Hendrickx, D. (2001). Pleidooi voor een grondbeleid. *Samenleving & Politiek, 8*(6), 29–33

Heynen, H. (2010). Belgium and the Netherlands: Two different ways of coping with the housing crisis, 1945–70. *Home Cultures, 7*(2), 159–177. https://doi.org/10.2752/175174210X12663437 526133

Lacoere, P. (2023). Van groei naar grens. Ruimteneutraliteit en bouwshift als doelstellingen voor duurzaam landgebruik. Gompel&Svacina

Lacoere, P., Hengstermann, A., Jeling, M., & Hartmann, T. (2023). Compensating downzoning. A comparative analysis of European compensation schemes in the light of Net Land Neutrality. *Planning Theory & Practice, 24*(2), 190–206. https://doi.org/10.1080/14649357.2023.2190152

Lacoere, P., & Leinfelder, H. (2020). Tijdloze ruimte. Planning voor onbepalde duur, of hoe de gewestplannen tot stand kwamen. *Ruimte, 12*(47), 12–17

Lacoere, P., & Leinfelder, H. (2022). Land oversupply. How rigid land-use planning and legal certainty hinder new policy for Flanders. *European Planning Studies, 31*(9), 1926–1948. https://doi.org/10.1080/09654313.2022.2148456

Pisman, A., Vanacker, S., Willems, P., Engelen, G., & Poelmans, L. (Eds.) (2018). Ruimterapport Vlaanderen (RURA): een ruimtelijke analyse van Vlaanderen / 2018. Departement Omgeving

Stad Gent (2018). Ruimte voor Gent. Retrieved October 11, from https://stad.gent/nl/wonen-bou wen/stadsvernieuwing/toekomstvisie-voor-stadsvernieuwing/ruimte-voor-gent#Publicaties

Shahab, S., Hartmann, T., & Jonkman, A. (2021). Strategies of municipal land policies: Housing development in Germany, Belgium, and Netherlands. *European Planning Studies, 29*(6), 1132–1150. https://doi.org/10.1080/09654313.2020.1817867

Vandermeer, M.-C., & Halleux, J.-M. (2017). Evaluation of the spatial and economic effectiveness of industrial land policies in northwest Europe. *European Planning Studies, 25*(8), 1454–1475. https://doi.org/10.1080/09654313.2017.1322042

Vigano, P., Cavalieri, C., & Barcelloni Corte, M. (Eds.) (2018). *The horizontal metropolis between urbanism and urbanization.* Springer

Vlaamse overheid (1997). Ruimtelijk Structuurplan Vlaanderen. Retrieved October 11, from https://omgeving.vlaanderen.be/sites/default/files/2022-01/RSV2011%20%282%29_0.pdf

Vlaamse overheid (2018). Strategische Visie Beleidsplan Ruimte Vlaanderen. Retrieved October 11, from https://www.vlaanderen.be/publicaties/beleidsplan-ruimte-vlaanderen-strategische-visie-geillustreerde-versie

Jean-Marie Halleux is professor at the University of Liege, where he teaches economic geography and spatial planning. His research interests are in the relationships between geography, urban economy and spatial planning. He has been involved in several European research programs (Interreg, COST, PUCA, JPI Urban Europe) where he has developed an expertise in the international comparison of land policies and planning systems. His comparative research focusses on planning culture, soft densification, land sobriety (no net land take), land value capture and the impact of industrial land on economic development.

Hans Leinfelder is professor in planning policy and vice dean for education at the faculty of architecture of KU Leuven. He is also member of the P.PUL research group at KU Leuven. In the past, he has been working at Ghent University and as policy maker and director at the spatial planning department of the Flemish Government. His research is on substantive aspects of urbanism and spatial planning in an urbanized countryside context (nature, agriculture, forestry, recreation, landscape). Second focus is on planning policy instruments as tools to translate policy options into (legal) frames for decision making on spatial development.

Land Policy in Czechia: When Underregulation Brings More Conflicts

Eliška Vejchodská

Abstract Land development in Czechia often proceeds without detailed land-use plans, leaving the decision on the type of development—whether the land is going to be used for housing, administrative offices, or retail buildings—to the discretion of the developer. This paper examines a site near Prague's Botanical Garden, at the edge of a residential area, where a developer has advocated for constructing a large retail centre resembling a hypermarket. This proposal conflicts with the preferences of local politicians and the public, who favour other types of development. The dispute has transformed the administrative procedures for development approval into arenas of conflict, resulting in prolonged underdevelopment of the site and substantial costs ineffectively spent by all stakeholders involved. The chapter explores how underregulation can lead to increasing conflicts, underscoring the need for more comprehensive planning policies.

1 Czech Understanding of "Land Policy"

The Czech language lacks a direct equivalent term for "land policy", encompassing public policies affecting the rights and duties of landowners in the pursuit of various political goals. When the term "land policy" was used within EU legislation,[1] it was officially translated into Czech as *pozemková politika*. This expression is much narrower than "land policy", however. The term *pozemková politika* is usually used in connection with the changes in land ownership initiated by the public sector, such as in connection to land ownership strategies of municipalities.[2] The Czech Senate

---

[1] https://eur-lex.europa.eu/legal-content/CS/ALL/?uri=CELEX%3A52004DC0686.

[2] Government resolution No. 292/2017, Strategický rámec Česká republika 2030. Available at: https://www.cr2030.cz/strategie/.

E. Vejchodská (✉)
Faculty of Social Sciences, Charles University, Prague, Czechia
e-mail: eliska.vejchodska@fsv.cuni.cz

© The Author(s) 2025   53
T. Hartmann et al. (eds.), *Land Policies in Europe*,
https://doi.org/10.1007/978-3-031-83725-8_4

uses this term as a component of agricultural policy solely.[3] Other uses of this term are related mainly to land reforms affecting land ownership in Czech history.

Even though the Czech language lacks an equivalent term for "land policy", various legally defined public policies exist within the family of land policy. Land policies are embedded within the legal framework through various acts under different ministries' purview.[4]

Of utmost significance in land policy matters is the Ministry of Regional Development, which is the central governmental body responsible for Planning and Building Act 183/2006 Coll. and newly 283/2021 Coll. taking legal force on 1 July 2024 *(Stavební zákon)*, Expropriation Act 184/2006 Coll. *(Zákon o vyvlastnění)*, and the preparation of a new act on the housing support. The latter is aimed at enhancing housing affordability for different social groups.

Another ministry involved is the Ministry of Agriculture, under which Land Consolidation Act 139/2002 Coll. *(Zákon o pozemkových úpravách)* operates. The act delineates the conditions for altering the spatial structure of property rights specifically within the scope of agricultural land, excluding urban land.

The Ministry of Environment represents a third critical ministry encompassing acts related to land policy instruments. For instance, it administers Environmental Impact Assessment (EIA) Act No. 100/2001 Coll. *(Zákon o posuzování vlivů na životní prostředí)*. The agenda of this act is highly relevant to the case studied in this chapter.

At the local level, Czech municipal self-government bodies implement land policies defined by law to pursue local interests. A Czech specificity is the large number of municipalities. With over 6000 municipalities for a population of 10 million inhabitants, Czechia has the smallest average municipal size among OECD countries (OECD, 2023). Because of municipal autonomy fragmentation, the set of land policy instruments available for municipalities is relatively narrow, and according to many land policy actors, it is not accommodated to the needs of the largest cities. The low availability of land policy instruments also has consequences for the use of the plot *K Pazderkám* within the capital city of Prague, selected as a case study.

2 The Mystery Behind the Plot '*K Pazderkám*', Prague

Between the popular Prague housing estate area of *Bohnice* on the one side and the Botanical Garden of Prague, integrated into the natural park area *Drahaň-Troja*, on the other side, there is an 11,500 m^2 plot (see Fig. 1). Currently, this plot is a wild green area of trees that started to grow uncontrolled several decades ago when it stopped serving as a construction yard for the development of the Bohnice housing estate. This plot is developable but is still waiting for its time to come.

---

[3] https://www.senat.cz/xqw/xervlet/pssenat/eurovoc?de=6833.

[4] Competencies of ministries are defined by competence law No. 2/1969 Coll. (kompetenční zákon).

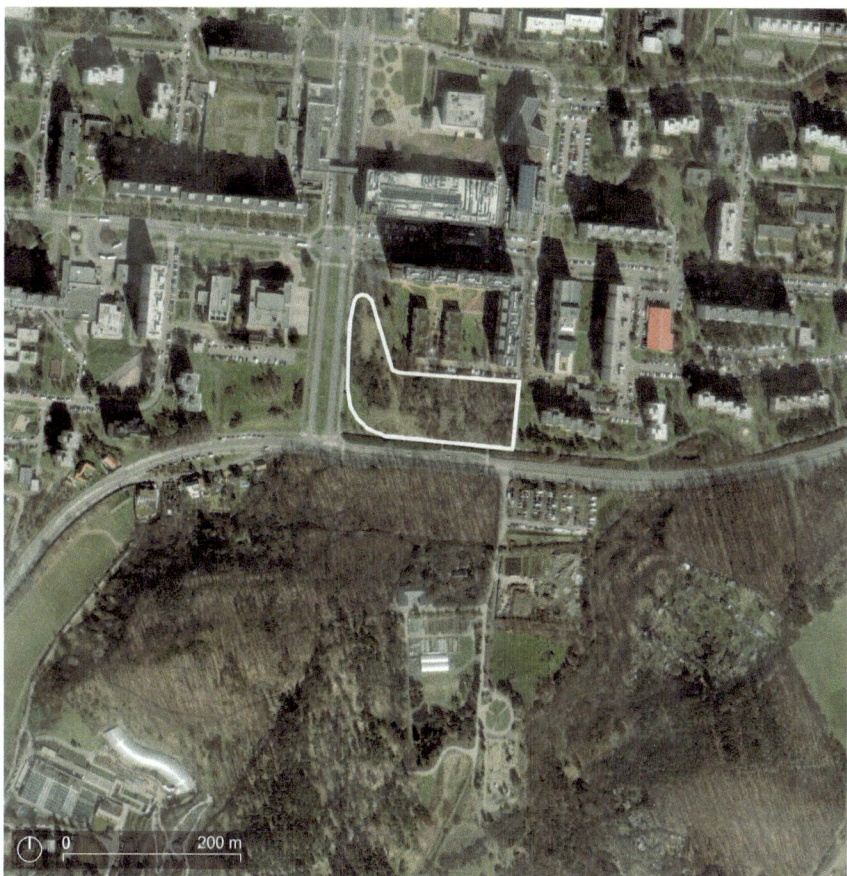

Fig. 1 K Pazderkám in Prague, Czechia (map: Kleiner, Jehling 2024, Aerial Imagery Provider © Esri, Maxar, Microsoft)

2.1 The General Context of the Location

The surrounding housing estate was constructed on the hill above the Vltava River in the seventies in response to the Czech baby boom and the pressing need to accommodate a large number of families. Although many planners in the nineties were concerned about the potential stigmatisation of these socialist housing developments for their architectural style, low quality and high number of apartment units, their inhabitants seem to have established a strong sense of attachment to these places, have made significant investments in their homes, thereby contributing to the preservation of their high value (Špaček, 2012).

The residents of these housing estates value the extensive green spaces surrounding the buildings and the ample provision of local public services and civic amenities within them. The Bohnice housing estate is well equipped with schools and preschools, a large healthcare centre, a public cultural centre, numerous sizeable grocery stores, and a shopping mall.

The value of local apartments has continuously increased hand in hand with the increasing demand for housing in Prague. The number of inhabitants of Prague has been rising almost continuously, surging from 1.18 million in 2000 to 1.36 million in 2022 (CZSO, n.d.a). Further increase in the population of Prague until 2050 is expected, as reflected in the new binding land-use plan of Prague under finalisation.[5] The influx of new inhabitants pushes the prices of the existing housing stock in Prague up and generates considerable demand for developing new housing projects.

The plot of land *K Pazderkám* is a precious plot for potential housing development thanks to its proximity to the natural park area, high standard of public services in the immediate vicinity, and good connection to the city centre via public transport. Nevertheless, the investor prefers to build a large retail centre here. Figure 1 shows the location of the plot *K Pazderkám*.

2.2   The Opportunities for the Plot K Pazderkám Set by the Land-Use Plan

The planning document specifying the development opportunities for the plot *K Pazderkám* is the land-use plan adopted by the city of Prague as a by-law. The land-use plan specifies the allowed types of development for the whole jurisdiction of the municipality and is legally binding. Its enactment represents the final opportunity for politicians to decide on future land development within developable areas.

Because land-use plans need to be enacted for the whole territory of the municipality, they usually do not go into detail in large cities. At the same time, municipalities, including the city of Prague, only sometimes enact locally specific binding zoning plans to delineate types of development further, thus often leaving the details of development to investors' discretion. This level of investor autonomy is rare even in planning systems considered neo-liberal, where, as Falleth et al. (2010) describe, detailed planning is often delegated to investors but still requires political approval. The Czech planning system offers even greater freedom, eliminating the need for political approval of investors' plans. Land-use plans enable landowners to secure planning permission from civil servants, provided their intention aligns with the often relatively broad conditions set by the land-use plan and meets all legal requirements (Vejchodská & Hendricks, 2023). The land-use plan does not guarantee that all land development intentions compatible with land-use plan can be realised. The possibility of development needs to be re-examined first. This examination also contains the point of view of environmental impact assessment (EIA), if the proposed development is large enough. EIA is essential for the plot *"K Pazderkám"* and will be discussed later.

---

[5] Metropolitní plán, 2023, available at https://plan.praha.eu/.

Since 1999, when the land-use plan of Prague was enacted, it has been amended many times. However, this plan became outdated with the enactment of a new Czech planning law in 2006. The 2006 planning law states that all land-use plans enacted before 2007 need to be adjusted to the requirements of the new law, and if not enacted until 31.12.2022, the old versions of land-use plans would lose their force (183/2006 Coll., §188, part 1). The city of Prague decided to develop a brand-new plan instead of adjusting the old one. The city council decided on the assignment for the new plan in 2013 (Praha, 2022). The new land-use plan was being developed for more than ten years. The whole process of creating a new land-use plan for a large city proved to be extremely difficult and time-consuming. The situation became stressful when it was clear that the new land-use plan could not be approved before the deadline set by law with the consequence of the opportunity for an unrestrained development coordinated solely by much less specific regional planning documentation. Eventually, the city of Prague gained an exception from the Ministry of Regional Development for being allowed to keep the outdated land-use plan in force even after 2022. In 2024, the new plan was expected to be finalised and approved in 2025.[6] This story of lengthy procedures of creating and approving a new land-use plan reflects the complex nature of establishing a binding land-use plan for a large city.

The plot *K Pazderkám* is defined as developable in both land-use plans—the old as well as the new one. The old one allows on a part of this plot generally mixed development, allowing the investors to develop multi-functional buildings or a combination of mono-functional buildings for housing, commercial, administrative, cultural, public facilities, sports and services, while maintaining multifunctionality of the area. It also specifies the intensity of such development—the maximum permissible development is 2.2 m^2 of floor area per 1 m^2 of land, including specific conditions for attaining a certain amount of greenery within this part (IPR Praha, 2023). Another part of this plot is supposed to fulfil the function of municipal greenery according to the old land-use plan.

The new land-use plan under finalisation changes the character of the regulative definition of development intensity by defining the maximum number of floors (for the plot *K Pazderkám* is relevant the regulation of up to 4 floors). The rest of the regulations stands as an accompanying text to the larger urban unit (in this case Bohnice housing estate) specifying the overall desired character of the broader area without going into the detailed specification of allowed land uses for partial pieces of land. According to the new text regulations, the plot *K Pazderkám*, as well as all other developable plots within the Bohnice housing estate, should become suitable for whichever of the following functions by the enactment of the new land-use plan: housing, retail and other civic amenities or administrative buildings (Praha, 2022).

---

[6] https://www.archiweb.cz/news/praha-chce-V-tomto-volebnim-obdobi-schvalit-zpozdeny-novy-uzemni-plan.

2.3 The Conflict Surrounding the Plot K Pazderkám

The plot *K Pazderkám* is not owned by a passive landowner waiting for better development prospects. Its landowner is an active entrepreneur employing available measures to obtain permission to construct. Commencing the formal process by applying for a derogation for more parking spaces in 2017, the investor has encountered many obstacles resulting in minimal progress in formal process even after many years (the case is described here to the situation of year 2024). How is this possible, given the favourable planning regulations pertaining to this plot?

The reason lies in the type of development that the investor has pushed forward: a large retail centre devoted to attracting considerable car traffic, thereby increasing noise and pollution in its vicinity. Furthermore, the project would add only negligible value to the already abundant shopping opportunities within the Bohnice housing estate for local inhabitants.

The contentious nature of the project has drawn the attention of residents and an informal association of active local people who called themselves *Proti plotu* (later formalised as an NGO) when the investment intention became public in 2017. Above all, *Proti plotu* has become an active opponent of the project.

If no other actor had raised complaints, the case would have perfectly matched the not-in-my-backyard (NIMBY) effect, describing residents typically opposing large commercial developments in their vicinity (Brown & Glanz, 2018). However, the project also elicited negative reactions from the city council, local district councils, and attracted the attention of other NGOs. Many highly respected personalities from the Czech scene expressed their opposition to the first version of the project by signing the petition initiated by *Proti plotu* in 2018.

In stark contrast to the investor's endeavour to develop a retail centre, the opposing side has advocated for an alternative vision. A survey conducted by the NGO *Proti plotu* revealed that residents prefer the development of low-rise apartment buildings in combination with homes for the elderly, a new preschool, sports or green areas. Only 3% of respondents expressed a preference for another shopping centre.[7]

The development of apartment buildings instead of a new retail centre on the plot *K Pazderkám* aligns with national priorities as well as current priorities of Prague. Low housing affordability is currently high on the political agenda, as real property prices more than doubled between 2010 and 2023 (CZSO, n.d.b). Many stakeholders within development consider that low housing affordability is caused to a great extent by lengthy approval processes lasting many years. Due to this belief, the new planning law was approved in 2021 with the vision to expedite the processes.

The case of plot *K Pazderkám* shows why development approval processes usually take so long in Czechia. The reason often lies in the low level of details of land-use plans, which prevents civil servants from conducting the administrative process without being embroiled in various conflicts between the interest of the land developer and the interests of other parties. The lengthy process incurs high costs for the developer related to planning, coordination, and project management, often requiring

---

[7] https://protiplotu.wordpress.com/2021/03/08/.

multiple redesigns and resubmissions of documentation prepared with great technical detail. Additionally, investor costs often increase due to the long-term possession of vacant land. The lengthy administrative process also brings high costs for public administration bodies, as officials must repeatedly address the same issues. The vagueness of land-use plans brings uncertainties into the administrative planning-permit approval phase by accumulating conflicts not solved during the political decision-making period (Lunardon, 2017).

Various parties attempt to obstruct or contest the administrative authority's decision concerning planning-permit application by highlighting procedural flaws or raising objections based on certain laws. It is not feasible to oppose projects based on urban design or the desirable type of development, thus aspects typical of the political decision-making process, as administrative authorities are not authorised to decide on subjective urbanism matters (Vejchodská & Hendricks, 2023).

As a result, the administrative procedure, while solving the substance of administrative matters, also needs to respond to objections raised by different stakeholders, leading to significant delays. The long and exhausting conflict around the plot *K Pazderkám* (see Fig. 2) has been in place for many years resulting in the plot remaining undeveloped without any resolution in sight.

According to Dawkins (2000), "land-use plan represents a collective contract between various land use interests and the local government" to decrease the costs of negotiations between the different parties affected. However, a too vague binding land-use plan does not effectively fulfil such an intention. At the same time, it curtails the ability of local politicians to strategically direct development, address the city's emerging challenges, and promote public interests.

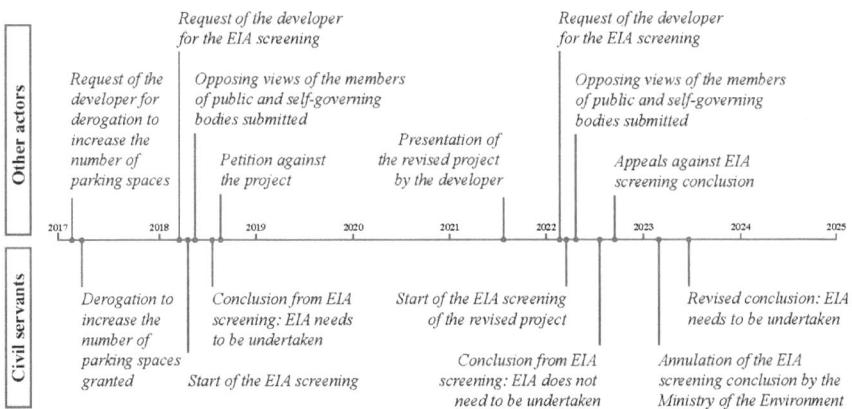

Fig. 2 Timeline of the process for the project K Pazderkám (*Source* authors)

3 Actors in Czech Land Policy

This section depicts the actions of various actors of Czech land policy in the context of the struggle for and against this project's realisation (for its timeline, see Fig. 2).

3.1 Private Landowner and Its Activities to Push Through the Development

The landowner of the plot *K Pazderkám* is a private company active in land development. In Czechia, ownership information for all plots of land is publicly available through the land cadastre, and anyone can access this information free of charge. The land cadastre also provides historical purchase prices of plots for a fee.

The private landowner in question acts on behalf of a company operating numerous large retail centres in Czechia and abroad. It is likely that these companies entered into a purchase agreement or development contract for this particular plot. Since the landowner initiated the approval process in 2017, the retail chain's name has been stated in the documentation title. This information became public as a result.

The proposed retail centre exceeded the limit of 6,000 m^2 of the built-up area, which triggered the requirement of screening the necessity for EIA according to the EIA Act N. 100/2001 Coll. Otherwise, the investor would enter the planning permit approval phase immediately.

The investor submitted the screening document to EIA to the city hall in February 2018. The document stated the main characteristics of the project and expected level of impact on various environmental components. The screening document was made public on the EIA information portal. Various stakeholders may have sent written views on the screening document to the competent authority within 30 days of the date of publication of the notification (Act N. 100/2001 Coll., §6, part 6).

The project raised massive negative attention from various actors, submitting their negative views. Three months later, the authority issued its conclusion from the screening phase deciding that the environmental impacts of the project should be thoroughly assessed according to the law (see Footnote 8).

Several years after the release of the screening phase conclusion, the investor did not undertake any further steps in pushing through the project, at least from the perspective of an external observer. In 2021, the investor requested the cancellation of the EIA conclusion for the original project[8] and presented a revised and renamed project with the title referring also to "a park with water elements". The public presentation intended to showcase developing a retail centre as a wonderful opportunity for this particular location and the city.

---

[8] https://portal.cenia.cz/eiasea/detail/EIA_PHA1066.

The revised project entered the EIA screening at the beginning of 2022.[9] Initially, the Prague authority responsible for the EIA process concluded that the project would not significantly affect the environment; thus, the full EIA report is unnecessary. After receiving appeals against the decision, the Ministry of Environment, acting as the appeal authority, cancelled the first decision and returned the case to the Prague authority for reconsideration. In March 2023, the Prague authority office decided that the project could potentially significantly impact the environment, necessitating the submission of the full EIA report.

The landowner's next intentions remain unclear. Will the company proceed with the full EIA report, or will it reconsider the entire project? Speculation is the only option at this point. In the following sections, we will delve deeper into the positions and interests of other relevant actors who have played a significant role in the story of this plot.

3.2 Members of the Public Fighting Against the Project

The EIA law offers an advantageous position for members of the public to express their views on a project subject to evaluation under the EIA law. Anyone can express his or her position. In contrast, under the planning law, only selected stakeholders, typically the landowners whose land directly borders the plot under the planning permit approval process, can express their positions.

In 2018, the EIA public authority received 552 reactions, mainly negative, which were considered relevant to be included in the conclusion from the screening phase. Most reactions were written by individuals living nearby the project who invariably expressed dissent. In 2022, the EIA public authority received an even more substantial influx of reactions, reaching 1765 individual submissions from the members of the public. These reactions usually built their critique on the increase in transport volumes due to the new retail centre, exacerbation of air pollution and noise levels within the vicinity of the retail centre, negative impact on public health and on the surrounding nature and landscape (see Footnotes 8 and 9).

In light of the public authority's final decision that the project must be assessed by a comprehensive EIA, all viewpoints need to be scrutinised in the EIA documentation, and if relevant, they need to be embedded as obligatory conditions integral to the final EIA conclusion.

3.3 Public Authorities and Land Policy

Diverse public actors possess the capacity to intervene in the planning or EIA process. These actors encompass administrative bodies as well as local self-government

---

[9] https://portal.cenia.cz/eiasea/detail/EIA_PHA1143.

bodies. Administrative bodies safeguard public interests within their respective domains based on the requirements of the law, while self-government bodies safeguard local interests based on deliberation.

The most important public actor significantly affecting the EIA process in 2018 by submitting its position towards the project under evaluation was the city of Prague as the self-government body through its expert organisation founded by Prague responsible for preparing new planning documents, preparing expert statements and coordinating the city's development, called Prague Institute of Planning and Development (IPR). According to their assessment in the first EIA process in 2018, the project did not align with the land-use plan because of the lack of public green spaces within the area owned by the investor and an encroachment upon the delineated zone as public greenery by the land-use plan (see Footnote 8).

In 2022, the city of Prague re-emerged as a significant actor in the EIA process, adopting even a notably more critical position. Its critique of the project was underpinned by many aspects connected to city urbanism. The city of Prague requested, among others, (i) an appraisal of the capacity of the existing retail network within the Bohnice housing estate relative to analogous urban enclaves in Prague. If the capacity proved sufficient in comparison to other housing estates, any escalation in transport volumes arising from the proposed retail centre would be considered by the city of Prague as highly problematic; (ii) the coordination of the project with the public transport development intentions of the city of Prague in the vicinity—with regard to the new tram line connecting the Bohnice housing estate with the nearby underground and the terminal of the new cableway connecting the northern part of the city via Bohnice housing estate with the Vltava river's opposite bank; (iii) the relocation of the retail centre's supply yard away from the vicinity of the cableway terminal and the tram line stop, to enable attaining a high standard of public space contiguous to this planned transport nexus where significant pedestrian movement is anticipated. The city of Prague notes that this request might not have any good solution and therefore questions the whole intention of developing a retail centre there.

The future development of the administrative process will show, if the critique raised by the city of Prague will effectively prevent the project from being realised. Municipalities as self-government bodies are allowed to present whichever standpoints and requirements in their statements. Administrative bodies assess their relevance in respect to the law.

4   Institutions of Czech Land Policy

The most important municipal planning document is the binding land-use plan, adopted by municipalities as a by-law. As stated above, it enables developers to apply for planning and building permission. Civil servants grant the planning and building permission if the intended development corresponds to the regulations defined by the land-use plan and other conditions set by law.

The procedure of planning permission approval allows various affected parties to submit nonbinding statements, meaning their input and opinions are considered but do not have the power to influence the final decision directly. Previous research shows (Feřtrová et al., 2013; Vejchodská & Hendricks, 2023) that apart from the formal ways of interference, municipal politicians also harness informal channels to exert influence over prospective land development. The existence of informal opportunities to influence administrative procedures stems from their dominance over civil servants providing decisions as administrative bodies, although administrative bodies should be politically independent. Administrative officers are city hall employees, and city councillors can affect their position. As a result, civil servants providing decisions as administrative bodies are often willing to comply with the wishes of local politicians in their decisions.

The overall ability of politicians to obstruct the approval process leads to developers' common willingness to negotiate with politicians about the form of development and voluntary financial contributions to the city even after the enactment of the binding land-use plan (Vejchodská & Hendricks, 2023).

This discussion shows how important informal institutions, like behavioural conventions of different actors within the planning process, are for the development in Czechia. In the following part of this section, we will further discuss only formal land policies which might be activated to make progress in the development of the plot *K Pazderkám* in a suitable manner for the city and its inhabitants.

4.1 Passive Land Policy via Land-Use Plan Provision

Through land-use plan provision, municipalities provide opportunities for landowners to build, yet the land-use plan does not impose a duty to build (Hengstermann, 2018). Some land-use plans are relatively specific concerning developmental type and character, while others leave the investor greater leeway in aligning their intentions. Subsequent stages in this process are left up to the activity of landowners. If they are interested in developing the land, they may apply for planning and building approval by civil servants as administrative bodies. The municipality does not actively push the development of the designated area any further.

Municipalities also utilise the passive land policy approach when numerous individuals own a development site. Czech planning law does not contain any instrument designed to facilitate the delineation of property boundaries according to the needs of future development, commonly known as land readjustment (Hong & Needham, 2007). The responsibility for the property layout changes rests in Czechia solely upon potential investors interested in developing it. These investors usually endeavour to buy up the parcels from individual owners. Only if they successfully consolidate the property ownership can they apply for a planning permit. Many areas that are earmarked for development by urban plans stay underdeveloped for this reason in Czechia.

Development of a plot is not subject to any obligatory provisions of public infrastructure or financial contributions to a city, as is relatively common in advanced economies (an overview of approaches can be found, e.g., in Muñoz Gielen & van der Krabben, 2019) apart from the provision of necessary technical and transport infrastructure. Typically, this includes the full provision of technical and transport infrastructure within the development area, sometimes also extending the capacity of technical and transport infrastructure beyond the area if the proposed development necessitates such extension (Vejchodská & Hendricks, 2023).

Many cities started collecting voluntary contributions from developers to partly cover investment costs of other types of public infrastructure, mainly public schools and preschools. Recently, many larger cities, including Prague (Praha, 2022), codified their own rules for voluntary contributions to enhance the transparency of the process of contributions and predictability of this cost burden for developers. As a result, land value capture (Alterman, 2012) has become more prevalent, allowing cities to recoup some of the increased land value generated by development projects.

The plot *K Pazderkám* has only one landowner, and therefore, its development is currently not disabled by the fragmented ownership structure. If there were no conflicting interests between the developer, on the one side, and the city and local inhabitants, on the other side, and if the development type did not potentially cause any significant harm to the environment as defined by the EIA law, the landowner could have agreed with the voluntary payment and smoothly proceed within planning and building permit approval phase.

4.2 Reactive Land Policy via Land Use Plan Change

The city-wide land-use plan is not rigid. It is subject to modifications in response to evolving municipal needs or when landowners apply for changes through formal channels. Decisions about land-use plan changes rest within the purview of municipal self-government bodies.

In the past, land-use plans have undergone revisions and provided landowners with additional building rights without mandating any reciprocal contributions from landowners (usually not removing existing development rights as this would necessitate compensation in certain circumstances). Recently, some cities, including Prague, embedded provisions for contributions associated with augmented building rights within their codified schemes of voluntary contributions based on private law. The contribution agreement has to be formalised prior to the change in the land-use plan.

The current proprietor of the plot *K Pazderkám* could have explored the opportunity, engaging in negotiations with the city government bodies about the nature of development alterations. However, this proprietor has not been proactive in this way.

The new planning law (283/2021 Coll.) also allows municipalities to activate obligatory in-kind or monetary landowners' contributions in return for an augmentation in building rights based on public law. The law does not specify any conditions on the quantum of contributions and their possible utilisation, therefore endowing autonomous self-government entities with discretionary powers in their application.

4.3 Active Land Policy—is it an Alternative for the Plot K Pazderkám?

Active land policy involving the purchase of land by municipalities to realise their development intentions is not a typical approach used by Czech cities. Only if land acquisition is necessary for the provision of large public infrastructure projects, they buy land. Municipalities have sold their land in the last decades instead of purchasing additional parcels. However, a paradigm shift has been observable. City representatives discuss opportunities to provide affordable housing on municipally owned land. Prague, probably the first Czech city, established its own development entity for that purpose, the Prague Development Company (PDS). This company does not purchase any land either.

Considering the hesitance of the city regarding land acquisitions, the purchase of the plot *K Pazderkám* from its current proprietor by the city of Prague, alternatively a land exchange, for enabling a non-conflicting use, appears unlikely as a costly approach to guard the public interest. On the other hand, this alternative cannot be dismissed, as historical precedence illustrates a case when the city of Prague did exchange land to avoid an undesirable development within a forest in the vicinity of the Botanical Garden, close to the plot *K Pazderkám*.[10]

5 Reflection

The plot *K Pazderkám* exemplifies a typical conflict surrounding a development intention. Analogous conflicts arise elsewhere when developers advance projects through the administrative phase that align with private interests while not adequately considering public interests and the visions of elected municipal representatives.

In the last decade, the lengthy bureaucratic procedures connected to development approvals started to be heavily criticised by land development investors, who perceive these procedures as obstructive, as well as by the public, which has blamed long approval procedures for high housing prices. This situation resulted in the codification of a novel planning law in 2021.

Revisions within the new planning law centred on the requirement of compliance with deadlines of administrative processes and on their simplification with the trust that this is the most critical problem of the previous law to solve. However, the new law does not solve one of the most significant reasons for long administrative

---

[10] https://www.archiweb.cz/n/domaci/ekocentrum-V-troji-nebude-praha-smeni-S-investorem-poz emek.

processes—that administrative procedures are overwhelmed with battles that should have been solved before. The problem seems to be rather underregulation, which stems from the lack of detailed planning.

The case *K Pazderkám* shows that a land-use plan that was too vague for the area led to the escalation of conflict between opposing parties. The investor has invested considerable resources in preparing various studies and documentation for the EIA process which turned out to be useless. Many years ago, the investor bought the land and has faced the burden of holding an expensive asset without utilisation for so long. He has a high stake here and upholds the investment idea of building a retail centre. On the other hand, the city and the public endeavour to impede the investor from obtaining necessary approvals, both investing a lot of effort too.

A potential remedy for the current Czech practice, which would require a change in the law, is the possibility for cities to combine a non-binding, preparatory land-use plan for the entire city area or for parts of the city where it would be reasonable, with locally specific, binding land-use plans. The city-wide preparatory plan can provide a broad overview with less detail than today, while the locally specific binding plans can offer more detailed regulations. This approach would allow solving all possible conflicts of interest during the preparation of the binding land-use plan without postponing them to administrative processes and make elected officials responsible for the final development character.

Not all vacant plots within Czech cities face analogous conflicts. Some developable plots stay vacant primarily due to the passivity of their landowners, who regard land as an asset for future use. Other prospective developable sites face a fragmented ownership structure that prevents their effective use for development. Also, within the Czech context, we can, therefore, refer to the situation in which land activated for development is scarce in comparison to the relative abundance of land defined by land-use plans for development (Hengstermann, 2018).

The development processes are slow in Czechia, but their pace must be fixed by more than bureaucratic process reform. The most significant causes reside elsewhere: the lack of appropriate instruments for land readjustment and the creation of detailed planning documentation. The Ministry of Regional Development has established an expert panel to set a new vision for planning law. We will see how successful this effort will be in fixing these issues.

References

Alterman, R. (2012). Land use regulations and property values: The "windfalls capture" idea revisited. In N. Brooks, K. Donaghy, & G. J. Knaap (Eds.), *The Oxford handbook of urban economics and planning* (pp. 730–755). Oxford University Press.

Brown, G., & Glanz, H. (2018). Identifying potential NIMBY and YIMBY effects in general land use planning and zoning. *Applied Geography, 99*, 1–11.

CZSO. (n.d.a). Obyvatelstvo (Inhabitants). Regional administration of the Czech Statistical Office in Prague. Retrieved July 23, 2024, from https://vdb.czso.cz/vdbvo2/faces/en/index.jsf?page=statistiky

CZSO. (n.d.b). Ceny bytů (Housing units prices), Czech Statistical Office. Retrieved July 31, 2023, from https://www.czso.cz/csu/czso/ceny_bytu

Dawkins, C. J. (2000). Transaction costs and the land use planning process. *Journal of Planning Literature, 14*(4), 507–518.

Falleth, E. I., Hanssen, G. S., & Saglie, I. L. (2010). Challenges to democracy in market-oriented urban planning in Norway. *European Planning Studies, 18*(5), 737–753.

Feřtrová, M., Špačková, P., & Ouředníček, M. (2013). Analýza aktérů a problémových aspektů rozhodování při nakládání s územím v suburbánních obcích. In M. Ouředníček & P. Špačková (Eds.), *Sub Urbs: Krajina, sídla a lidé* (pp. 234–256). Academia.

Hengstermann, A. (2018). Building obligations in Switzerland: Overcoming the passivity of plan implementation. In J.-D. Gerber, T. Hartmann, & A. Hengstermann (Eds.), *Instruments of land policy* (pp. 175–188). Routledge.

Hong, Y., & Needham, B. (Eds.). (2007). Analysing land readjustment: Economics, law, and collective action. Lincoln Institute of Land Policy.

IPR Praha. (2023). Regulativy funkčního a prostorového uspořádání území hlavního města Prahy. Retrieved July 19, 2023, from https://www.praha.eu/jnp/cz/o_meste/magistrat/odbory/odbor_uzemniho_rozvoje/uzemni_planovani/uzemni_plan/index.html

Lunardon, W. (2017). Urban planning—Czech republic in the European context. *Urbanismus a Územní Rozvoj, 20*(1), 28–30.

Muñoz Gielen, D. & van der Krabben, E. (Eds.). (2019). Public infrastructure, private finance: Developer obligations and responsibilities. Routledge.

OECD. (2023). Subnational governments in OECD countries: Key data. Retrieved July 23, 2024, from https://search.oecd.org/regional/multi-level-governance/NUANCIER-2023_rev_compressed.pdf

Praha. (2022). Usnesení zastupitelstva hl. m. Prahy č. 33/8 ze dne 27.1.2022 ke schválení Metodiky spoluúčasti investorů na rozvoji území hl. m. Prahy. Retrieved August 23, 2023, from https://zastupitelstvo.praha.eu/ina/SeznamList.aspx?aid=3&sid=0&pid=1&mid=1,81

Špaček, O. (2012). Česká panelová sídliště: Faktory stability a budoucího vývoje. *Sociologický Časopis/czech Sociological Review, 48*(5), 965–988.

Vejchodská, E., & Hendricks, A. (2023). Munich's developer obligations as a legal transplant to the Czech institutional context. *Town Planning Review, 94*(2), 193–214. https://doi.org/10.3828/tpr.2021.28

Eliška Vejchodská is an Associate Professor at the Faculty of Social Sciences, Charles University in Prague, Czechia, within the Department of Public and Social Policy. Her research primarily centres on the instruments of environmental policy, land policy, and on urban governance.

Land Policy in England: Maintaining Equilibrium in a Contested System

Janet Askew

Abstract Land policy in England is delivered through laws and regulations, statutory instruments, sectoral, environmental, and planning policies at national and local levels. The use of land is controlled and implemented by an increasingly complex and bureaucratic adaptive planning system. Successive governments in the past 40 years have sought to simplify the system. The case study is based on The Wirral, a peninsular on the edge of Liverpool's conurbation. Housing development is proposed across seven un-related, protected green field sites. The case illustrates several fundamental misunderstandings, including that of national and local policy. These aim to prevent urban sprawl and protect green land, where locally there is sufficient brownfield land to meet demand for new housing. The applicant ignores the validity of public involvement in decision-making, as well as the procedures for submission of planning applications. Following rejection of the housing sites based on long-standing national and emerging local policy, an appeal is summarily dismissed by a government inspector. Numerous questions about how and where housing should be provided are raised, but this story also seeks to demonstrate that the planning system does not need full-scale reform to ensure it is working well.

1 Land Policy in England

In England, there is no reference to the term 'land policy'. Using the term to compare approaches to spatial planning and policies regarding the use of land across different countries in Europe results from different approaches to land law (Common Law in England, Roman Law in continental Europe). Comparing planning systems must also consider that the form of a planning system is determined by the 'underlying social model, values and cultural assumptions' that have evolved through history, and these shape the way planning is performed (Nadin & Stead, 2008). In other words, planning methods and systems are not just 'an outcome of… laws and instruments,

J. Askew (✉)
European Council of Spatial Planners-Le Conseil Européen des Urbanistes (ECTP-CEU),
Brussels, Belgium
e-mail: janetaskew@townplanning.eu

© The Author(s) 2025
T. Hartmann et al. (eds.), *Land Policies in Europe*,
https://doi.org/10.1007/978-3-031-83725-8_5

but a demonstration of a planning culture' in the country of study, and there will be a different approach to terminology and definition of planning between countries according to the legal families (Nadin et al., 2024).

Land policy (policies about the use of land) in England[1] is delivered via a variety of laws and regulations (land and planning), statutory instruments, sectoral, environmental, and planning policies at national and local levels. There is no national plan in England, nor is there a strategic level of planning outside London. The use of land is both encouraged and restricted by this combination of laws and policies, and its implementation is undertaken at the local level through the granting of planning permission, which is required for all land uses, however small. The planner at national or local level is charged with surveying, analysing, and planning for proposed changes in land use, whilst politicians at all levels make the ultimate decisions on how the land will be used. In the English system, many stakeholders are involved in this process, including public and private sectors, civil society, communities, residents, individuals, developers, sectoral interests including government agencies, NGOs, and environmental charities, some of whom are designated as statutory consultees in the planning process.

The planning system in the UK, nuanced for England, is an adaptive system, distinguished by two aspects: the nationalisation of development rights (Town and Country Planning Act, 1947) resulting in a discretionary system of controlling development which depends on negotiation; and a policy-led approach from statutory (but not legally binding) development plans alongside government frameworks and statements. The introduction of both in 1947 was controversial, perceived by landowners as an introduction of controls over privately-owned land, as up until this point they were involved in regulating their own land, largely to protect their investments and enhance the value. The new system compounded shifts in the state-market-civil relations which has continued for over 70 years (Lloyd, 2015). The regulation of all land created a dichotomy between public and private landowners and has arguably resulted in a constant attack on planning in the UK, with continuous calls for reform and de-regulation, emphasised in recent years by the influence on government of neo-liberal lobbyists and developers (Lloyd, 2015).

Despite reforms, very little has changed in principle during the past 70 years regarding the principles of planning law and regulation, although local plans have become more policy-oriented, omitting rigid land use designations. Calls for simplification by private interests have eroded the planning system, not necessarily for the better, making it more complex than it used to be, creating the exact opposite of a simpler system (Clifford et al., 2018). Instead, it is more bureaucratic, cumbersome, slow, unpopular, and expensive and 'the failures laid at the door of planning are often failures of the state' (Sturzaker & Sykes, 2023). The answer is a properly resourced planning system which is central to addressing any nation's needs (RTPI, 2023). It is the view of many planners in the UK that instead of embarking on more reform

---

[1] This case study refers only to England and not to the UK. The devolved administrations of the UK - England, Scotland, Wales, and Northern Ireland - have different legislation and although the principles of the statutes are similar, the laws and policies differ slightly. (Sheppard, A., et al. (2017).

and deregulation, a wiser approach would recognise the importance of stability and the positive potential of planning (Sturzaker & Sykes, 2023).

The Wirral case study demonstrates the tension between the municipal authorities, the government, and a private landowner, as well as the public and civil society, all of whom are stakeholders in the act of changing the use of land. The biggest single issue in England is housing, which is at crisis point, and its development emphasises the tension between state-market-civil relations.

2   Understanding English Land and Planning Policy: A Case Study of Housing on the Wirral in North-West England.

The case study concerns a proposal for housing on the Wirral, a small rectangular peninsula in north-west England, located in the west of the Liverpool metropolitan area (see Fig. 1). It has an area of $264 \, km^2$, bound on three sides by sea and river coasts, with a population of approximately 400,000. The division between the two halves of the Wirral is palpable, with the eastern side comprising two towns, Birkenhead and Wallasey, once home to prosperous docklands, now one of 20% of the most deprived areas in England (MCHLG, 2019); whilst the opposite is true of neighbourhoods on the western side, which are so wealthy as to be some of the most prosperous in England. Housing demand in the west of the Wirral comes from its attractive landscape and coast, semi-rural villages, good commuter transport links—rail, ferry, and road tunnels serving destinations all over the peninsula, with direct access to Liverpool.

The case study is an outline application lodged with Wirral Borough Council by Leverhulme Estates for multiple housing sites spread throughout the prosperous, more rural areas in the centre of the Wirral on designated 'green belt' land. The green belt is a national statutory land designation, applying only to the land surrounding 15 urban areas in England, the aim of which is to protect the countryside around cities from urban sprawl (MHCLG, 2024). Leverhulme Estates owns 2000 ha of land on the Wirral, including farms, housing, industrial and retail premises. It has long-term plans to eventually build 8000 houses across 400 ha on its green land (Wirral Globe, 2023). This proposal for seven sites is a test application, with intention to bring more sites forward later. The applicant relies on its history of 'great place-making', building on its reputation of (nineteenth century model village) Port Sunlight and other historic and aesthetic villages across the Wirral (Leverhulme Design Charter, 2022), offering to adopt historic townscapes to justify the new building.

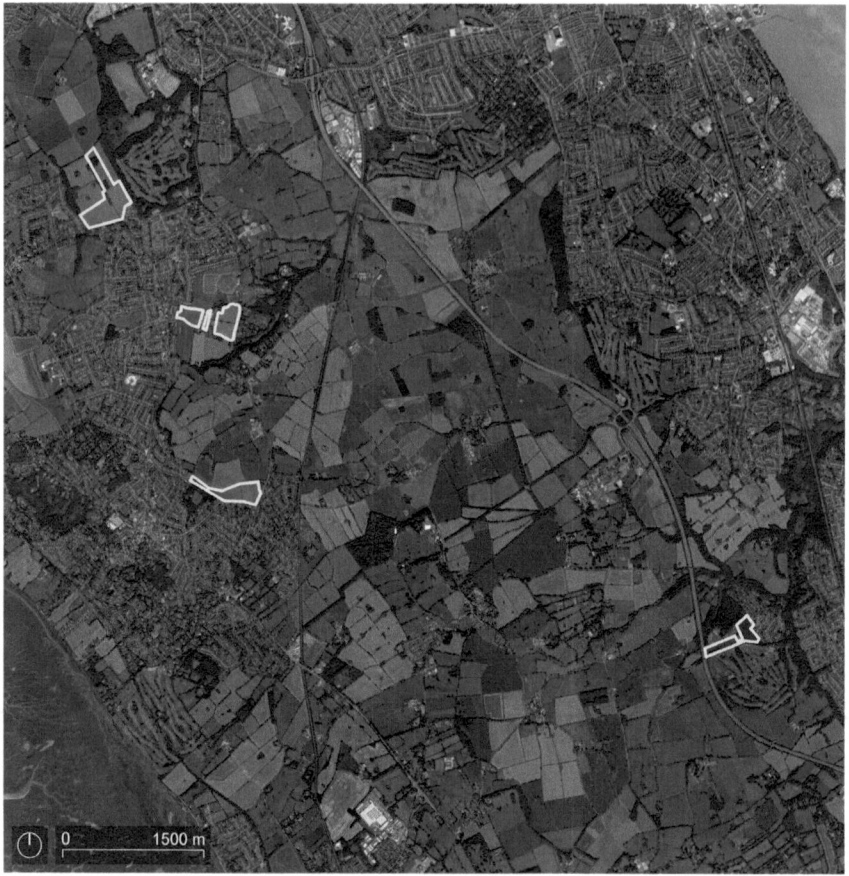

Fig. 1 Aerial view of the scattered sites proposed for housing in The Wirral Green Belt (map: Kleiner, Jehling 2024, Aerial Imagery Provider © Esri, Maxar, Microsoft)

2.1 Land Valuation and the Impact on Housing

There are no national goals (as in some European countries) to reduce landtake in general, and land protection is achieved via appropriate policies, such as the latest policy for environmental protection of land, which aims to protect 30% of land for biodiversity by 2030 (Wildlife and Countryside Link, 2023), as well as planning policies such as green belt designations. The biggest issue in this case study is the role of local and national planning policy, which advocates very strong protection of certain land which is designated as 'green belt'. The policy has been stable since 1955 but is currently coming under constant attack by market forces (eg. thinktanks, such as Centre for Cities, Adam Smith Institute, Policy Exchange), who lobby governments, and academics such as Cheshire (2019). Housebuilders campaign hard for building on green field lands due to ease of building, which creates opportunities for higher profits. (Foye & Shepherd, 2023). Semi-rural land adjacent to urban areas commands the highest prices per hectare for housing and with planning permission, houses can be sold at a premium, even though building costs are higher due to infrastructure provision. The debate continues with the 2024 government introducing legislation to allow more building on green belt land, calling it 'grey belt'. This is land which, although in the green belt, might not have any green or rural attributes, although this fundamentally misunderstands the concept of green belt in relation to urban sprawl. However, the land that is the subject of the Leverhulme Estates is largely green field land.

Promoters of building on rural land ignore the fact that this will not resolve the affordable housing crisis. Research shows that land-hungry, low density, detached single dwellings are mainly proposed on green field land across England—in 2023, only 5% of the housing built on green belt was social housing (CPRE, 2023). On the prosperous and attractive west side of the Wirral, land for housing attracts the highest values on the peninsula and sites with planning permission offer high returns for the landowner. Here, developers argue that providing the requisite number of affordable or social houses would not be viable due to land prices. Local and national policy states clearly that building on brownfield land should take priority over green fields. The latter should only be used when sufficient brown field land is not available for the number of houses needed. Leverhulme Estates argues that the demand for housing in the Wirral can only be met by building on green fields. However, the local plan for Wirral identifies a large amount of brownfield land for housing in the existing urban areas of Wirral.

Viability of a development is not a material consideration in planning law in England. The only exception to this is over the negotiation of planning gain agreements, where the planning authority seeks contributions of social housing, infrastructure, and other mitigation, at which point the developer might claim that their scheme is not viable if they comply with demands for gains. An understanding of development economics, yield and viability is an essential requirement for professional planners in the English system to enable successful negotiations in the public interest.

2.2 Planning Permission in the English Planning System

The English planning system depends on hierarchical policies at the national and local levels to guide development and land use.

England has a *'National Planning Policy Framework'* (MHCLG, 2024) which contains very general policies relating to all land uses, alongside a 'presumption in favour of sustainable development'.[2] National planning policy (NPPF) is accompanied by a raft of rural land classifications in England, including national parks,

---

[2] Major infrastructure – power stations, wind farms, power lines, major railway lines, sewage and waste, water, flood defences etc. are dealt with differently through a nationally-determined

areas of outstanding natural beauty, sites of special scientific interest, green belts, agricultural land and biodiversity, flood risk vulnerability areas, all protected and delivered through planning policy at the local level in co-ordination with sectoral semi-governmental and government agencies. Rural land, if not protected by any of the above policies is covered by a national generic 'rural land restraint policy' (MHCLG, 2024) which strictly restricts the use of rural land to certain uses, mainly to prevent urban or random sprawl.

The primary material consideration for controlling land use and deciding on what should be built is the statutory 'development plan' or 'local plan', which is not legally-binding, produced by local planning authorities at the urban or rural district or city-wide level. It is a largely policy-based document, which does designate some land for certain uses such as housing, and significantly for this case study, it defines the green belt boundaries. Once this plan has been through numerous stages of public consultation and revision, it is examined for 'soundness' in a government inquiry, eventually to be 'adopted' as a statutory plan. Whilst 80% of the 324 local planning authorities in England have an 'adopted' plan, 50% of which are being updated (slowly), some 64 local planning authorities have no plan at all (Williams et al., 2023). Each local plan must demonstrate a five-year housing land supply which invites controversy from communities and developers alike and contributes to the delay in housing provision.

Much smaller areas, such as villages or a neighbourhood within a city, can prepare a 'neighbourhood plan', in which land is allocated for various uses. They are not compulsory (unlike the local plan) and whilst there are thousands of districts in England which are eligible to make a neighbourhood plan, by no means do all of them have one. It is estimated that there are 2800 plans under preparation in England (Parker et al., 2023). Introduced partly to overcome the opposition to new housing, it was thought that if communities themselves could designate housing land, it might remove the objections (Parker et al., 2023). Wealthier communities are more likely to initiate a neighbourhood plan than deprived communities, often to ensure that land is protected from housing development (Sturzaker & Sykes, 2023).

All the plans mentioned, at national, local and neighbourhood levels must conform to each other. Statutory plans can be challenged at any time during the process, especially if they are out of date or have not been reviewed.

Although the local plan, once adopted, is the primary material consideration, the discretionary planning system of England allows for planning applications at any time, whether or not an up-to-date local plan exists. Applications for development are considered on merit against all national and local policies. All regulations pertaining to the process of how to develop or change the use of land, including those minor uses for which planning permission is granted automatically by government, (permitted

---

parliament decision-making process which differs from the norm, and they are determined by a government agency (not the municipality).

development) are laid out in statutory instruments (GPDO, 2015; UCO, 2020 [as amended]).[3]

The main difference between this indicative (adaptive) system and the imperative system of legally binding plans is the point at which the decision is made to build (Faludi, 1987; Lacoere & Leinfelder, 2022). In England, that point is when planning permission is granted, subject to:

- no conflict with 'approved', 'adopted' or 'made' policy. This is applied hierarchically with the statutory development plan first, other plans, and then any material considerations; and
- any conditions or agreements (i.e. for land value capture) that the local planning authority (i.e. local council) wishes to impose.

Planners play an important role in negotiation, another distinguishing feature of the English planning system. This requires professional expertise, lifting planners from being mere bureaucrats (or technocrats).

Planning applications allow for a developer to seek 'outline permission' to establish the principle of a particular use on a piece of land. Leverhulme Estates adopted this approach. If agreed, details are submitted with ever-expanding requirements for more technical information (impact studies, public consultation, options and amendments, economic viability). Thus, applying for planning permission is a very lengthy and expensive process (a fee is payable). Successive governments responding to developers' demands to seek de-regulation have somehow resulted in yet more regulation (Clifford et al., 2018).

The way that national policies, local plans, and complex laws and regulations interact with a discretionary planning system is pertinent for this case study, which demonstrates how national policy, in particular, green belt designation, impacts the provision for new housing at district level. Being a test case for the applicant, an unusual feature is that the applicant applied for 788 houses on (seven) multiple non-adjacent sites across the Wirral (see Fig. 2). This is rare but allowed in law. All the sites are in the Wirral Green Belt. This multiple site approach substantially reduced the likelihood of receiving planning permission for any of them. A planning application spread over scattered or non-adjacent sites must be dealt with as one application—if one site is refused, they all would be. This approach also created more controversy with policy makers and the public, owing to the wide spread of sites on protected land.

Following submission, the options open to the local planning authority (politicians are the decision-makers on the advice of their expert officers who are qualified planners) include refusal or approval, with reasons or conditions respectively. In Wirral, after a long campaign by the public, civil society and other stakeholders, and with the support of local politicians, all sites were refused planning permission

---

[3] It is in this area of planning that the minor alterations have largely taken place. The government increases the number of applications which can be classed as 'permitted development', most recently in the field of housing, the conversion of industrial and commercial premises into residential use, without full planning permission.

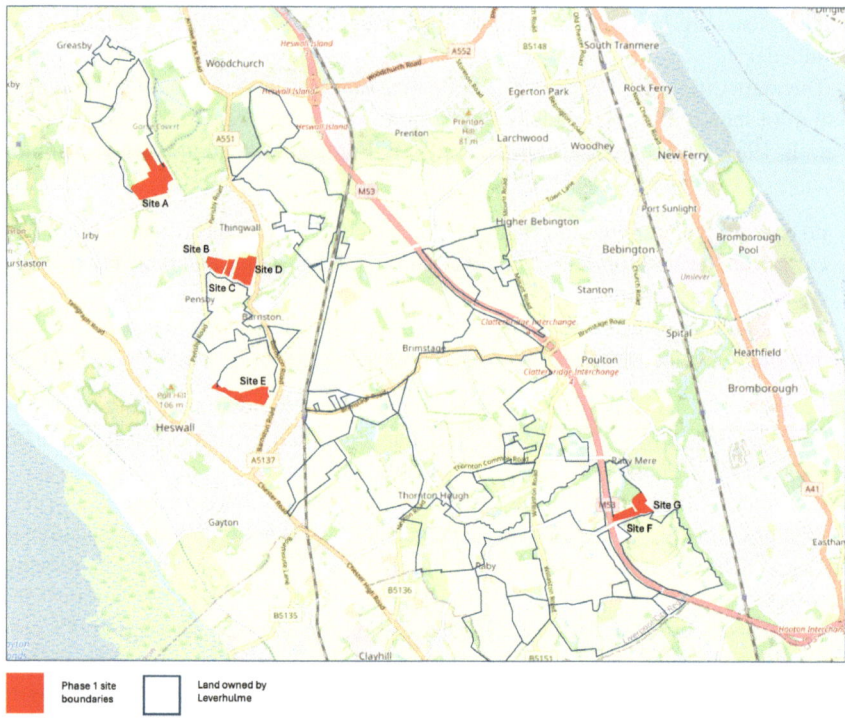

Fig. 2 Seven sites for housing (A–G) located beyond the boundaries of villages on the Wirral and in the Wirral Green Belt (*Source* Author; map data from OpenStreetMap 2024-08-23, under the Open Database Licence version 1.0 (the ODbL). Site boundaries from Planning Inspectorate Decision Letter Reference APP/W4325/W/22/3313729, 31 January 2023, data by Alan Baxter Ltd., p. 6)

by the local authority on the grounds that development of the sites would conflict with green belt policies; and that the application was 'premature' in advance of the adoption of the new local plan, which would include a five-year housing land supply.

These reasons for refusal were upheld on appeal, the decision concluding that "the very special circumstances necessary to justify the developments (in green belt) do not exist and there are clear reasons for refusing the developments proposed."[4]

Furthermore, in an unusually strongly-worded decision, a full award of legal costs was justified to be paid to the local authority due to.

> … the appellant's unreasonable behaviour in pursuing and maintaining the residential appeals …. which … have no realistic prospect of succeeding. (PINS 2023)

It is an interesting case study because:

- It was unusual for the submission of seven non-adjacent scattered sites under one application.

---

[4] *(PINS decision (13 September 2023) can be found at:* https://acp.planninginspectorate.gov.uk/ViewCase.aspx?caseid=3313729*).*

- The applicant is famously the nineteenth century philanthropic landowner and developer of one of the first model villages in the UK (Port Sunlight) which is admired the world over.
- The applicant has taken a very cavalier and market-led, profit-led approach to building houses on green belt land, with no regard for its protection under local and national policy.

Adding to the lengthy negotiations in the passage of a planning application, the UK is well-known for its experimental attempts to capture land value, with many different approaches over the years. Known as the 'world's former laboratory of betterment capture instruments' (Alterman, 2012), successive governments have investigated the best ways to tax increases in land value since 1932. Different ways of capturing surplus value have been tested, and from 1971, surplus value was used to offset the impact of development. This is negotiated at length during the passage of a planning application, and whilst it cannot influence whether permission is granted (i.e. it is not possible to 'buy' a planning permission), planning permission will not be granted until the planning agreement is signed. Latterly, a two-pronged approach combining certainty (akin to a land tax) and negotiation has been employed. Requests for gains must be judged 'reasonable' in law and must avoid corroding the neutrality of the planning system. In addition to the negotiated agreements for affordable housing, the local authority can charge an infrastructure levy for large-scale construction projects to provide public services. A planning application can be permitted subject to the signing of a legal agreement, but strictly speaking, they are not part of the legal permission to build.

To control the use of land, and how it is used, planning permission is always granted subject to 'conditions' which relate directly to the site. These can restrict certain uses of the land; phasing of the building; landscaping; density, height and building materials; drainage—all referencing national regulations. The legitimacy of planning conditions can be challenged and determined by the courts, but conditions on use of land are usually accepted in most cases.

In the Wirral application, a range of very general 'gains' was offered, in which the applicant suggests that for each of the seven sites, there would be 30% 'affordable housing', green infrastructure, biodiversity net gain, contributions to environmental matters, green cycle ways, public open space etc. Assuming all sites received planning permission, all would have had to be negotiated separately, and this would have held up building on any site—another reason why submitting seven sites in one application was unwise, and the wording of the application regarding 'planning gain' renders it meaningless.

2.3   Reducing Land Consumption and the Housing Crisis in England—Green Land or Brownfield Land?

The applicant's reason for applying for housing in these locations, making a departure from national and local policy, is that it will help to address the severe housing shortage in England. Social housing owned by councils was sold off after the Housing Act, 1980, local authorities ceased to build more, considerably reducing the provision of houses for rent. By allowing social renters to purchase their houses at a discounted price, it removed obligation on the state to maintain social housing stock (cost reductions), and the necessity for the state to provide subsidised housing because the private rented sector would fill the gap. This did not happen. Successive governments since 1980 have not successfully addressed housing shortages, and the onus is on local plans to demonstrate a five-year housing land supply. Government estimates 300,000 new houses need to be built each year (in perpetuity) to meet demand. The actual annual build figures fall far short of this—in 2022, 172,000 houses were completed (ONS, 2022); only 8400 social houses were built in 2022–23 (New Economics Foundation, 2024). Other factors, including inflation, high interest rates, high rents, and scarcity contribute to the housing shortage.

There is a controversial debate about shortage of land. Whilst there are long-standing 'brownfield first' national and local policies, there remain obvious barriers to building on brownfield land (e.g. cost of de-contamination, land assembly). Those in favour of restriction and protection of countryside allege that England has sufficient brownfield land (urban regeneration) for housing (CPRE, 2023), and that urban sprawl is unsustainable. Others are of the view that urban fringe and rural land is over-protected and needs release for housing. In both scenarios, historic and current national planning policies are blamed for the lack of housing, particularly green belt policy. An English peculiarity is that there is considerable opposition to building housing, contributing to long delays in building out new infrastructure and housing, including urban extensions, new settlements, urban fringe infill, regeneration of brownfield sites in rural and urban areas. The residents of rural areas or those who live in the green belt do not want neighbours or new houses to despoil the landscape. The case study illustrates this in the number of objections recorded to the building of houses by Leverhulme Estates.

Whilst Leverhulme Estates uses the housing shortage to justify its application, it pays no respect to the green belt policy; nor to the fact that Wirral Borough Council can demonstrate in its local plan a five-year housing land supply on brownfield land in the east of the Wirral, without any incursion into the green belt. The council is committed to building up to 12,000 houses on derelict, unused land, and a planning application is already in the pipeline. Although strong policy protects green belts, developers continue to advocate for building on England's 15 green belts, whilst countryside and sustainability campaigners argue for its retention. Nothing is more controversial in planning debates in England at present.

Green belts were included in the reform of the planning system in the Town and Country Planning Act, 1947, and local authorities in 14 locations around cities were instructed by Government via Circular 42/55 in 1955 to define boundaries of green belts whose purpose was to:

- Check the unrestricted sprawl of large built-up areas
- Prevent neighbouring towns from merging
- Safeguard the countryside from encroachment
- Preserve the setting and special character of historic towns
- Assist urban regeneration by encouraging the recycling of derelict and other urban land.

England no longer has strategic planning authorities, meaning that the building of large new areas of housing, such as urban extensions requires co-operation across local authorities to define and protect from development rural areas through green belt and rural restraint policies (which are separate to green belt policies)[5] (Rankl et al., 2023). There are currently 84,500 houses proposed on land to be released from the green belt in adopted local plans in England, and 123,400 proposed in emerging local plans, amounting to just under 208,000 houses in total (CPRE, 2023). The opposition from Wirral residents to release of green belt land for housing shows the extent of the controversy. The Liverpool/Manchester/West Yorkshire Green Belt (in which the Wirral lies) lost 17,000 acres in seven years. (CPRE, 2019).

3   Actors in English Land Policy

In any major development, there are numerous actors involved in the process—state, market and civil. The case study reveals the role of both public and private stakeholders, and organised civil society, of whom those who advocate for protection of green belt and environmental areas are the strongest.

3.1   Private Actors and Land Ownership

The ownership of land is in the public domain, held by the Land Registry, which holds information about 88% of the land mass in England and Wales. It was established in 1862, and since 1925 it has been compulsory to register all titles of land. The public can seek information about ownership, although information must be paid for. Irrespective of who owns the land, the local and neighbourhood plans allocate land for various developments. The local plan invites comments from all stakeholders

---

[5] Green belt policies only apply to the 15 locations across England to prevent urban sprawl. Beyond urban areas, there prevails 'rural land restraint policy' which restricts building in rural areas to certain rural uses, the policies for which are contained in the National Planning Policy Framework (NPPF) for England.

during its passage, and landowners are likely to participate, either to promote their own land or object to any designations.

In the case of the Wirral housing sites, the applicant was the sole landowner. A legal aspect of planning law in England is that an applicant does not need to own the land when they apply for planning permission. The requirement is only to notify, not seek permission from, the landowner. Major housebuilders traditionally bank land into their own ownership, sometimes for many years, by buying it at agricultural (or lower) value and then waiting until the values have increased, and planning permission can be obtained to build. The Competitions and Markets Authority (2023) has analysed that 11 of the major housebuilders in mainland Britain are sitting on 1.17 million land plots for housing, over half of which constitute land banked for the future, and the remainder have planning permission which is not being built out. This number implies that planning permissions are being granted. A new power (LURA Act, 2023) enables authorities to refuse any application from a developer with a history of slow build-out rates. Housebuilders say that they will reduce the number of houses released onto the market for the near future so they can control their declining profits. In their report, the CMA (2023) reports that a 'complex and unpredictable planning system and the limitations of speculative private development' are responsible for the lack of delivery of new homes (Edgar, 2024). Planners argue that the lack of resources in municipal planning departments contributes to delay, where planning is not given priority due to massive austerity cuts in local government grants since 2010.

A common occurrence is the use of 'options' on land. The acquisition of land specifically for housing occurs when a developer approaches a landowner (in rural areas or in the green belt for example) and the landowner commits to maintain the land for the developer on condition that they pay an annual sum for this, including an assurance to apply for planning permission within a set period (e.g. up to 10 years). The aim is to buy the land at lower cost thus lowering the risk in the event of planning permission never being forthcoming.

The UK's concept of freehold and leasehold property rights is controversial at present, with calls for amendment to land law. A freeholder owns the land and the building on it; whereas a leaseholder owns only the land, but not the building. In this case, the owner of the building must pay a fee to the freeholder, and the freeholds are being sold on, creating higher costs for the leaseholders including annual rises written into contracts for new housing.

All land, public or private, is subject to the same planning laws, and planning permission must be sought for all developments, whoever owns the land. There is no special dispensation for either private landowners, nor for a government or state or institutional owner of land.

3.2 Public Actors and Land Policy

During the preparation of a local plan, public consultation forms an important part of it, and all stakeholders are eligible to support or object to proposals therein. Once a plan has been through this process, it is published as the 'adopted' plan for the period 10–15 years. However, if, during the life of a plan, an unexpected windfall site comes onto the market, such as land owned and no longer needed by a public authority (e.g. former defence, hospital, or government land), discussions will be held between public authorities to determine its future use. Despite the statutory nature of the plan, the discretionary, adaptive planning permission allows for any applications at any time which must be considered on merit. In the case of brownfield land coming to market, this gives the local authority an opportunity to engage in discussions and negotiations with the landowner to discuss the use of the land, especially if it offers housing opportunities in rural and urban areas. Owing to the state of public finances and shortage of public sector planners, much of the work of the planning application will be done by the landowner, public or private, although a (national) design code is now mandatory upon local planning authorities for large-scale developments.

Compulsory purchase powers exist, which might be used for access or infrastructure, but the public finances only allow this in special circumstances. Active land policy, whereby local authorities intervene directly in the market through purchase and servicing of land is unusual in England, due to the state of public finances. Where social housing is the preferred use, the most likely approach is that this will be provided on the back of a private development (through 'Section 106 agreements'); or a government funded 'housing association' will be involved to build the houses. There is pressure for local authorities to build houses (this largely ceased in 1980), but local authorities are once again starting to build social houses (small numbers to date) (Clifford et al., 2024).

In all cases, a refusal of planning permission for the development by the local planning authority can result in an appeal. Appeals are determined by a semi-independent government body, the Planning Inspectorate, which answers to the appropriate Secretary of State, who has the power to 'call in' an application for decision. If an appeal is dismissed, and there are procedural anomalies, a judicial review can be made to the high court. In the Wirral case, following refusal of the outline application on green belt and prematurity grounds, Leverhulme Estates has challenged the substantive reasons for refusal at appeal (rejected); and then again in a high court appeal on procedural grounds, which was also thrown out by the judge. They brought their application forward early to try to 'beat' the plan. In the case of the latter, the correct approach for any large landowner is to drive their influence or desires for site uses through the local plan process, something which Leverhulme Estates had done (Smulian, 2023).

Regarding public sector actors, Clifford et al. (2024) suggest that spatial planning is at a crossroads, with government reform undermining the traditional vision of state-employed planners making decisions about development in a unified public interest. Nearly half of UK planners are now employed in the private sector, with complex inter-relations between the sectors including supplying outsourced services to local authorities struggling with centrally-imposed budget cuts.

3.3   Civil Society/Third Sector Actors

The influential role of civil society organisations is a distinguishing feature of the UK, some of which buy and manage land and buildings to control development. In England, the third sector, or civil society plays a powerful role in protecting various interests, especially environmental. It has been suggested that because of the discretionary system, such civil bodies cannot trust the planning system to protect important environmental and historic environments. These powerful organisations have large memberships and financial support from the public. Two of the largest and most active are the National Trust (NT) and the Royal Society for the Protection of Birds (RSPB), both of whom own and manage large areas of protected land, and in the case of the NT, historic buildings, and villages, lobbying hard to protect these national assets (Cullingworth & Nadin, 2006). Upon receipt of a planning application, all these stakeholders and many more besides, along with the public, must be consulted (statutorily) for prescribed periods of time.

4   Institutions of English Land Policy

The planning system in the UK is highly centralised, adaptive, and discretionary, but this does not result in a laissez-faire system. A variety of instruments exist for the control of development including conditions; planning gain agreements; tariffs; time limits; enforcement, stop notices; requirement to build in accordance with the plan (Levelling Up and Regeneration Act 2023 (LURA). The courts provide stability to protect land policy in the UK. Policy is largely respected and appeals allow for argument to test local and central government policy. Planning law is constantly under review in England, but not always to the benefit of the public interest.

4.1   Land Policies for the Wirral Sites

The land policies for change of use of land are contained in national and local policies. The NPPF (MHCLG, 2021) lays down policies for all land, including in the case of restraint, outlining exemptions to the policy. Numerous national sectoral bodies (see above) can insist on restrictions on land-use—e.g. environmental agencies, national parks, in areas of outstanding natural beauty; agriculture and forestry restraints; and green belt designation. The latter is the most important for the case study. Monitoring the environment and taking enforcement action for any illegal activity are crucial elements in land policy, posing punitive sanctions, including re-building after illegal demolition, or imprisonment in some cases.

4.2 Passive Land Policy

The government lays down principles for planning. All regulations are centralised in England, more so under the new LURA Act (2023) which introduces national development management policies; and local government's policies must comply with the national framework. Interpretation by decision-makers relies on court decisions and precedents.

There is no land re-adjustment in England; the concept is not used. Some instruments exist to invite a more active land policy—compulsory purchase orders, for example, but owing to the chronic state of local government finance this is seldom used, except perhaps for major infrastructure. If local authorities own part of the land for a major development, they will be an active player in the negotiations especially regarding land assembly. Local authorities are often passive actors, waiting for developers to bring forward land for development either in compliance with a local plan or unexpectedly.

4.3 Reactive Land Policy

Whilst land might be designated for alternative uses, the actual development depends upon the local planning authority reacting to a developer coming forward with a proposal, such as happened in the Wirral. Due to the flexibility in the system, anyone can apply for planning permission for a development, irrespective of ownership, and irrespective of the allocation of land in a local plan. If a plan is up-to-date and recently 'adopted' by the council with approval by government, and land is protected, especially green belt, then planning permission is unlikely to be forthcoming. Developers should make these judgements in advance of embarking on a controversial site. The decision by Leverhulme Estates to pursue the application was considered unwise (by the judge in the high court appeal) on land protected by long-standing policy.

There is a change in the relationship between state and market in England. A complex marriage of state and market is emerging, whereby the state sometimes acts as a private company, mostly with the sale of land assets for profit. More qualified planners work for private consultants than for local authorities (RTPI, 2023), and authorities are outsourcing planning functions to the private sector, because the capacity to deal with planning applications and plans is reduced, so the use of private consultants, even global outsourcing companies, is an expensive answer to hard-pressed authorities. There is no doubt that this is a threat to local democracy, sometimes known as 'consultocracy' (Wargent et al., 2019). Since 2010, private landowners and developers have had a greater influence on planning in England through powerful lobbying of politicians by right wing think-tanks and big developers seeking to free up the market, against what they see as an inefficient and over-regulated planning system. A recent report by Adam Smith Institute calls for the 'height, width, space and density' of housing to be de-regulated, a potentially disastrous reduction in standards for new housing (ASI, 2023). Many applications for housing are made on green field sites which are not necessarily green belt, some of which gain planning permission, either by the local authority due to a need to provide housing or gained on appeal to the government.

Public consultation is compulsory by law in the UK. Although the local authority oversees this, increasingly developers manage and pay for it, creating concerns about their methods and vested interests.

4.4 Active Land Policy

The real-term cut of 40% in government funding to local authorities between 2010 and 2022, with several already declaring bankruptcy by May 2024 (Birmingham City Council for example), (Harris, 2024), prevents them being pro-active on development sites. Active land policy whereby the local government funds and facilitates a new housing development is unusual in the UK. Hard up local authorities also sell their own land for development and can name preferred developers or encourage consortia for large-scale developments. Section 106 agreements for planning gain allows local authorities to require a percentage of social/affordable housing to be delivered; or a financial contribution can be sought for social housing elsewhere, usually commissioned from a social housing provider (housing association). Value capture for public services can also be gained via a tariff on developments.

5 Reflection

The term 'land policy' is not used in the United Kingdom. Those aspects of the land management paradigm—land tenure, land value, land use, and land development—are not integrated in the English system (Williamson et al., 2009), despite their role in driving important support for sustainable development (Enemark, 2007). However, the dichotomies and tensions regarding land administration are demonstrated through a case study of a planning application for multiple housing developments in the Wirral, in North-West England.

In particular, the case study shows how the adaptive planning system of the UK has worked in the public interest for 75 years with stable policies, demonstrating that it is not a laissez-faire system. There is flexibility in the English system, which allows for modification in the use of land in response to changing circumstances, and this means that the discretionary planning system of the UK works well, within which qualified town and country planners have an important professional role to play. However, the influence of the private sector is increasing with calls for de-regulation of planning and a sympathetic response from government following intense lobbying, especially from the house-building industry (Foye et al., 2023). The solutions to the housing crisis in England lie beyond changes to planning policy and relate to fiscal measures, rent control, second homes regulations, better quality housing, land costs etc., but governments, at least since 1979, have called for planning reform. What would help is to halt the declining investment in state-managed planning, as local planning authorities are under pressure and short-staffed. Constant reform does not help, since professional planners have to apply new rules, some of which (changing planning permission rules for example for change of use from office to residential with lower standards imposed (Clifford et al., 2024) have resulted in seriously poor quality housing. In rural areas, the lack of strategic plans makes ad hoc developments more likely, especially as developers are banking land with hope for a future where rural and edge city land will be released for development, particularly in areas of rural restraint and release of green belt. However, national and local policy, after it has been 'adopted' following correct procedures, remains strong as a basis for determining appropriate land uses. In all cases, especially in relation to rural and conserved landscapes, powerful civil society organisations influence decision-making, and the public are extensively engaged in decision-making.

Looking at land value and administration issues, it is an important part of English planning that landowners do not necessarily have more power by dint of owning the land. In so many cases, private landowners believe that they have rights over the use of their own land, but because development rights were nationalised in 1947, all use and re-use of buildings and land requires planning permission. The value of land is determined by the granting of planning permission, which in turn determines the quality of the development. Land values, viability and profitability are largely separate from the planning system although the negotiation of land value capture through its many different instruments for doing so, requires professional planners to understand development costs and profits. In general, land value capture instruments assist in achieving infrastructure to enable development in the public interest, for the good of the community, often with protracted negotiations. In the case study in Wirral, the developers were aware of the necessity to negotiate some 'planning gain' on the back of developing such high quality, high value developments, but their approach was naïve and unconvincing. Approaches to land value capture remain controversial, and the numerous experiments in the taxation of increased value remain open to further experimentation in England.

These conclusions display a confidence in the English planning system which does not always exist amongst developers or the public at present. This case study shows that public, civil society, and politicians were at one in objecting to the development, based on national green belt policy, which has been in place since 1955. In this case, the five-year housing land supply could be met within the district where there are vast tracts of derelict former industrial land, provided the national and local policy reinforces this ambition.

Aspects of the case study demonstrate how acute the housing crisis is. It begs the question by the public (especially young people for whom housing is a distant dream), as well as by developers (who profit most from building on green field sites), as to whether the protection of so much green land is viable for the future. Building on brownfield land is prioritised but sustainable solutions for housing on other land should be sought. The absence of strategic planning across local authority boundaries is an impediment to good planning for housing, especially in areas where green belt occupies a large proportion of their land. There is a dichotomy in which the public objects to certain developments whilst at the same time thinking that the planning system is failing them. For some it is, and the challenge for planners in England is to persuade people that planning is in the public interest, but government reforms are not particularly helpful. The acquiescent relationship between the state and the market is potentially damaging for good planning and land policy in England, but intervention by civil society and the public seeks to balance this in the public interest.

References

Alterman, R. (2012). Land use regulations and property values: The 'Windfalls Capture' idea revisited. In N. Brooks, K. Donaghy, & G. J. Knaap (Eds.), *The Oxford handbook of urban economics and planning* (pp. 755–786). Oxford University Press.

ARUP. (2020, January 9). Wirral council empty homes report. Retrieved August 21, 2024, from https://www.wirral.gov.uk/files/h3-wirral-empty-homes-summary-scoping-task-1-report-2020.pdf

Birmingham City Council & British Waterways. (2002, April 1). City centre canal corridor. Development framework. Retrieved August 21, 2024, from https://www.birmingham.gov.uk/cityce ntrecanalcorridor

Booth, P. (2007). The control of discretion: Planning and the common-law tradition. *Planning Theory, 6*(2), 127–145. https://doi.org/10.1177/1473095207077585

Cheshire, P. (2019). Urban containment and housing affordability. Newgeography

Clifford, B., Ferm, J., Livingstone, N., & Canelas, P. (2018). Assessing the impacts of extending permitted development rights to office-to-residential change of use in England. RICS. Retrieved August 21, 2024, from https://www.rics.org/content/dam/ricsglobal/documents/to-be-sorted/assessing-the-impacts-of-extending-permitted-development-rights-to-office-to-residential-cha nge-of-use-in-england-rics.pdf

Clifford, B., Gunn, S., Inch, A., Schoneboom, A., Slade, J., Tait, M., & Vigar, G. (2024). The future for planners. In *Commercialisation, professionalism and the public interest in the UK*. Bristol University Press.

CPRE. (2019). *North-West Green Belt*. Campaign to protect rural England. Retrieved August 21, 2024, from https://www.cpre.org.uk/wp-content/uploads/2019/11/North_West_factsheet_2018. pdf

CPRE. (2023). *State of the Green Belt 2023: A vision for the 21st century*. CPRE Countryside Charity. Retrieved August 21, 2024, from https://www.cpre.org.uk/wp-content/uploads/2023/08/State-of-the-Green-Belt-2023-online.pdf

Competition and Markets Authority. (2023). *Housebuilding market study*. CMA. Retrieved August 21, 2024, from https://www.gov.uk/cma-cases/housebuilding-market-study

Crook, T., Henneberry, J., & Whitehead, C. (2016). *Planning gain: Providing Infrastructure and affordable housing*. Wiley-Blackwell and RICS.

Cullingworth, B., & Nadin, V. (2006). *Town and country planning in the UK (14th Ed.)*. Routledge.

Department for Environment Food and Rural Affairs (DEFRA). (2023, December 9). *Delivering 30by30 on land in England*. Retrieved August 21, 2024, from https://assets.publishing.service. gov.uk/media/65807a5e23b70a000d234b5d/Delivering_30by30_on_land_in_England.pdf

Department for Levelling Up, Housing and Communities. (2023). *Levelling Up and Regeneration (LURA) Act*. Retrieved March 5, 2024, from https://commonslibrary.parliament.uk/research-bri efings/cbp-9689/#:~:text=Planning%20reforms%20(England)&text=The%202023%20Act% 20will%20also,carried%20it%20out%20unreasonably%20slowly

Edgar, L. (2024, February 26). Unpredictable planning system contributing to housing under-delivery. The Planner. https://www.theplanner.co.uk/2024/02/26/unpredictable-planning-sys tem-contributing-housing-under-delivery?check_logged_in=1

Enemark, S. (2007). Integrated land-use management for sustainable development. In *Joint F.I.G. Commission 3, UN-ECE CHLM WPLA Workshop on informal settlements, Souino, Greece*, March 28–31, 2006. Retrieved August 21, 2024, from https://unece.org/sites/default/files/datast ore/fileadmin/DAM/hlm/sessions/docs2007/presentations/Enemark68thsession.pdf

Foye, C., & Shepherd, E. (2023). *Why have the volume housebuilders been so profitable? The power of volume housebuilders and what this tells us about housing supply, the market and the state. UK Collaborative Centre for Housing Evidence*. Retrieved August 21, 2024, from https://hou singevidence.ac.uk/publications/why-have-the-volume-housebuilders-been-so-profitable/

GDPO (2015). HM Government: The Town and Country Planning (General Permitted Development (England) Order 2015 (as amended) , London. Available at https://www.legislation.gov.uk/uksi/2015/596/contents

Faludi, A. (1987). *A decision-centred view of environmental planning*. Pergamon Press

Harris, J. (2024, January 14). One by one, England's councils are going bankrupt - and nobody in Westminster wants to talk about it. *Guardian*. https://www.theguardian.com/commentisfree/2024/jan/14/englands-councils-bankrupt-westminster?CMP=share_btn_url

Lacoere, P., & Leinfelder, H. (2022). Land oversupply. How rigid land-use planning and legal certainty hinder new policy for Flanders. *European Planning Studies, 31*(9), 1926–1948. https://doi.org/10.1080/09654313.2022.2148456

Letwin, O. (2018). *Independent review of build-out*. MHCLG, HMSO. Retrieved August 21, 2024, from https://assets.publishing.service.gov.uk/government/uploads/system/uploads/attachment_data/file/752124/Letwin_review_web_version.pdf

Leverhulme (2022, May 1). *Leverhulme Design Charter*. Leverhulme. Retrieved August 21, 2024, from https://programmeofficers.co.uk/Wirral/CoreDocuments/CD2/CD02.8%20-%20D esign%20Charter%20v5%20low%20res.pdf

Liverpool Echo (2023). https://www.liverpoolecho.co.uk/news/liverpool-news/laughter-controver sial-developer-said-care-26928013

Lloyd, M.G. (2015). *Provenance, planning and new parameters*. University of Amsterdam

LURA Act (2023). HM Government: Levelling-up and Regeneration Act 2023, London Available at https://www.legislation.gov.uk/ukpga/2023/55

McClements, D., & Hausenloy, J. (2023). *Cooped up. Quantifying the costs of housing restrictions*. ASI Research. Retrieved August 21, 2024, from https://static1.squarespace.com/static/56eddd e762cd9413e151ac92/t/65c2518698cfe71ca5997107/1707233675548/Cooped+Up+FINAL.pdf

MCHLG (2019), English Indices of Deprivation 2019 (IoD2019) London, MCHLG

MHCLG (2024) *National Planning Policy Framework*, London https://assets.publishing.service.gov.uk/media/67aafe8f3b41f783cca46251/NPPF_December_2024.pd

Nadin, V., & Stead, D. (2008). European spatial planning systems, social models and learning. *disP—The Planning Review, 44*(172), 35–47. https://doi.org/10.1080/02513625.2008.10557001

Nadin, V., Cotella, G., & Schmitt, P. (2024). *Spatial planning systems in Europe: Comparisons and trajectories*. Edward Elgar.

New Economics Foundation (NEF) (2024) Press Release: Government must build ten times more social homes by 2027/28 to meet housing targets, London Available at: https://neweconomics.org/2024/10/government-must-build-ten-times-more-social-homes-a-year-by-2027-28-to-meet-housing-targets

Office of National Statistics. (2019). *Urban green spaces raise nearby house prices by an average of £2500*. ONS. Retrieved August 21, 2024, from https://www.ons.gov.uk/economy/enviro nmentalaccounts/articles/urbangreenspacesraisenearbyhousepricesbyanaverageof2500/2019-10-14#:~:text=Homes%20close%20to%20parks%2C%20gardens,Statistics%20(ONS)%20a nalysis%20reveals

Office of National Statistics. (2022). *House building data, UK: financial year ending March 2022 update*. ONS. Retrieved August 21, 2024, from https://www.gov.uk/government/statis tics/house-building-data-uk-financial-year-ending-march-2022-update

Parker, G., Wargent, M., Salter, K., & Yuille, A. (2023). Neighbourhood planning in England: A decade of institutional learning. *Progress in Planning, 174*, 100749. https://doi.org/10.1016/j. progress.2023.100749

Rankl, F., Barton, C., & Carthew, H. (2023, December 15). Green Belt. House of commons library research briefing. Retrieved August 21, 2024, from https://researchbriefings.files.parliament.uk/ documents/SN00934/SN00934.pdf

RTPI (2023, July 14). *House of Commons' planning reform recommendations sensible and necessary*. Royal Town Planning Institute. Retrieved August 21, 2024, from https://www.rtpi. org.uk/news/2023/july/house-of-commons-planning-reform-recommendations-sensible-and-necessary/

Sheppard, A., Peel, D., Ritchie, H., & Berry, S. (2017). *The essential guide to planning law*. Policy Press.

Smulian, M. (2023, December 7). Leverhulme Estates to bring legal challenge of plans for 788 homes. Local Government Lawyer https://www.localgovernmentlawyer.co.uk/planning/401-planning-news/55838-leverhulme-estate-to-bring-legal-challenge-over-refusal-of-plans-for-788-homes#

Sturzaker J., & Sykes O. (2023). *Planning in a failing state. Reforming Spatial Governance in England*. Policy Press.

UCO (2020). HM Government: The Town and Country Planning (Use Classes) (Amendment) Regulations 2020, London Available at https://www.legislation.gov.uk/uksi/2020/757

Wargent, M., Parker, G., & Street, E. (2019). Public-private entanglements: Consultant use by local planning authorities in England. *European Planning Studies, 28*(1), 192–210. https://doi.org/ 10.1080/09654313.2019.1677565

Wildlife and Countryside Link (2023), 30 x 30 in England 2023 Progress Report, London. Access at https://wcl.org.uk/assets/uploads/img/files/WCL_2023_Progress_Report_on_30x30_in_Eng land_1.pdf

Williamson, I., Enemark, S., & Rajabifard, A. (2009). *Land administration for sustainable development*. ESRI Press Academic.

Wirral Globe (2023) Available at https://www.wirralglobe.co.uk/news/23960248.leverhulme-takes-controversial-housing-plans-high-court/

Janet Askew is a chartered town planner in the UK, immediate past president of the European Council of Spatial Planners and former president of the Royal Town Planning Institute. She has worked in all four nations of the UK, including heading a large planning school in Bristol, England, and currently she is a Visiting Professor of Planning Law in Ulster University, Belfast. Her main area of research is in legal and regulatory regimes and planning processes. Her current activities include membership of the World Economic Forum's Davos Baukultur Alliance and in 2024, she was named as a 'Woman of Influence' by The Planner journal.

Land Policy in Finland: Public Land Development Still Standing Strong, Even on Brownfield Land

Eero Valtonen and Heidi Falkenbach

Abstract Finland is a country where planning and other land policy are quite strictly separated both in the legislation and in the practice. Within this context, the public land development approach, where the municipality acquires the land, approves the land use plans, services the land with infrastructure and sells/leases the serviced building plots for private development, has been used as the standard approach for the development of previously undeveloped land. During the recent decades, this approach has also been increasingly employed to redevelop large previously developed sites. This chapter discusses the redevelopment of Penttilänranta in Joensuu, Eastern Finland. This redevelopment case has taken place on an old sawmill site. The case illustrates how the Finnish institutional framework, with the strong reliance on the public land development approach, offers limited alternatives for municipalities to facilitate the development by other approaches than public land development if the owner of the site is uninterested in developing the site. The analysis of the estimated costs and incomes of the Penttilänranta project plan highlights how these kinds of brownfield redevelopment projects increase the need for more comprehensive debate and discussion on the financial risk-taking by the municipalities with the public land development approach.

1   Land Policy in Finland

In this book, land policy is defined as 'the sum of governmental interventions that influence the rights and obligations of landowners, usually in pursuit of social, economic, or ecological political aims', combining the elements of both planning and other land policy (referred to in Finnish as *maankäyttöpolitiikka*). In Finland, planning (i.e. the policy of determining how the land can be used, *kaavoitus*) and

E. Valtonen (✉)
The University of Manchester, Manchester, UK
e-mail: eero.valtonen@manchester.ac.uk

H. Falkenbach
Aalto University, Espoo, Finland
e-mail: heidi.falkenbach@aalto.fi

T. Hartmann et al. (eds.), *Land Policies in Europe*,
https://doi.org/10.1007/978-3-031-83725-8_6

other land policy (i.e. other measures with which the municipality advances its spatial aims, *maapolitiikka*) are often treated as separate. This separation is inherent in the Land Use and Building Act (LUBA), the main legislation regulating land policy. It is also present in practice, where the two parts of land policy are often administratively separated and are carried out by individuals with different educational backgrounds. Whilst LUBA regulates planning relatively explicitly, in terms of both process and content, other land policy is regulated much more generically and vaguely. Thus, municipalities can quite freely set the aims of their other land policy and the implementation of the policy through instrument choice.

Land policy in Finland is mostly formulated and implemented on the municipal level due to the characteristics of the Finnish planning framework and the strong political and fiscal autonomy of municipalities. The Finnish land use planning framework is inherently plan-led, consisting of several layers of spatial plans. The state government only provides national land use guidelines but no spatial plan. Below the state level, there is a regional plan (*maakuntakaava*) prepared and approved by the regional council and two main layers of plans prepared and approved by the municipality: a local master plan (*yleiskaava*) and a local detailed plan (*asemakaava*). All these plans are legally binding land use plans. However, their legal effects are different. The only legal effect of a regional plan and a local master plan is that they must be taken into account as guidance when a lower-tier plan is prepared.[1]

The legal effect of a local detailed plan is that a building permit must be given to a development project that is in conformance with the local detailed plan and must be refused if it is not.[2] In practice, no significant development in Finland can happen without a local detailed plan allowing the development. The municipalities have the independent right to initiate, prepare and approve local detailed plans, whereas other stakeholders (including landowners and other levels of government) can partake in the process through participatory processes or through plan appeals as defined in LUBA. Notably, the right to appeal only concerns plan approvals. The municipal decisions of not initiating the planning process for a particular area or not completing the planning process to the stage of plan approval are not decisions that can be appealed. Along with the extensive rights to control development in their jurisdiction, municipalities also have responsibilities. An approval of a local detailed plan makes the municipality responsible for implementing the plan-related infrastructure and public areas, such as streets and parks.[3]

---

[1] A local master plan can also have the legal effect of defining the conditions for a building permit if it has been specifically mentioned in that plan. As this plan is approved by the municipality, it does not affect the fact that land policy in Finland is mostly formulated and implemented on the municipal level.

[2] LUBA allows the municipality to discretionarily approve minor deviations from the local detailed plan. In addition to being in conformance with the local detailed plan, a development project must meet several generic technical requirements (provided in LUBA), applying to all buildings developed in Finland, to receive a building permit.

[3] According to LUBA, these infrastructures must be provided when the implementation of the private uses in the plan makes their provision necessary. However, the municipality does not have to provide these if it would be unreasonable in terms of the financial conditions for the municipality.

In the other land policy, the municipalities advance their autonomously set aims. These typically include: (1) Actions to encourage plan implementation, as the planning framework only provides the landowner the right, but not the responsibility to implement the planned development, (2) actions to ensure cost recovery or value capture, to enable financing of the plan-related infrastructure and public areas, and (3) actions to promote social and environmental aims, such as provision of social housing. The Finnish land policy toolkit offers a broad variety of instruments enabling both passive and very active land policy approaches (Krigsholm et al., 2022).

In particular, it provides for public land development (Valtonen et al., 2017a), in which the municipality acquires the land to be developed, prepares and approves the land use plans, services the land with infrastructure and finally sells or leases the serviced building plots to (private) developers. Indeed, the public land development approach has been widely used for the development of undeveloped (greenfield) land and increasingly used in (brownfield) redevelopment contexts, as well (Krigsholm et al., 2022; Lönnroth et al., 2024; Valtonen et al., 2017a; Viitanen et al., 2003).[4] The case we discuss in this chapter illustrates the dilemmas, tensions, and policy debates when this approach is applied to brownfield redevelopment.

2 Redevelopment of Penttilänranta in Joensuu

The Penttilänranta case is a redevelopment project of a 33-ha vacant sawmill site located close to the town centre of the municipality of Joensuu in Eastern Finland (see Fig. 1). According to Statistics Finland (2024), the number of households in Joensuu at the end of 2023 was around 43,000 and in the urban agglomeration, whose central municipality is Joensuu, around 67,000.[5] In 1985, the number of households in the agglomeration was around 46,000, of which just above 25,000 were attributed to Joensuu. In the recent decades, both Joensuu and the agglomeration have been growing with an average annual growth rate of 1.4% for Joensuu and 0.5% for the rest of the agglomeration, but since 2016, the latter—excluding Joensuu—has seen the number of households decline in every year but one. This leaves the majority of the recent growth accounted to Joensuu, while its surroundings are shrinking.

Within this context of constant but rather modest growth in the number of households, with Joensuu outpacing the rest of the urban agglomeration, the municipality of Joensuu had to deal with the land use problem of how to facilitate the redevelopment of the vacant and derelict sawmill site in the prime waterfront location close to the town centre. As the site had been in the sawmill use for over 100 years, it was generally assumed that the land would be contaminated and that any development of the area would require remediation actions.

---

[4] Unfortunately, accurate statistics are not available on the extent to which the public land development approach is used in Finland.

[5] Due to the nature of the statistics, they can slightly underreport the number of households as they do not include homeless households or separate different households sharing a dwelling unit.

Fig. 1 Penttilänranta in Joensuu, Finland (map: Kleiner, Jehling 2024, Aerial Imagery Provider © Esri, Maxar)

The land was owned by Bonvesta, a subsidiary of the global forest industry corporation UPM. After the sawmill had closed in the late 1980s, the municipality had made several attempts to initiate the redevelopment of the site over a period of 20 years. As these negotiations with the landowner had not resulted in any concrete decisions to develop the site, in 2008, the municipality purchased the whole site, with the liabilities related to the potential contamination, for the purchase price of one euro. In the project plan the municipality prepared in 2008 after acquiring the site, the cost of remediating the contaminated land to allow redevelopment of the area was estimated to be 21.2 million EUR in the 2008 price level (City of Joensuu, 2008). Thus, the real acquisition price of the site could be argued to be 21.2 million EUR instead of one euro, as Bonvesta (and eventually UPM) could probably have been made responsible for the costs of the land remediation.

Table 1 The estimated costs and incomes for the municipality in the Penttilänranta case. Source: City of Joensuu (2008)

Costs	Million EUR (in 2008 price level)
Land remediation	21.2
Technical infrastructure	12.8
Service infrastructure	3.7
Incomes	
Building plot sales	34.2

In the project plan, the municipality set an aim to develop around 175,000 m^2 of mostly new residential floorspace with some retail and office floorspace. This target was set to meet the population target of around 22,00–33,00 inhabitants set in a master plan covering a broader neighbourhood, which Penttilänranta is part of. The estimated completion time in the project plan was 25 years and included several local detailed plans[6] to be prepared and approved to allow redevelopment of different parts of the site. The costs of remediating and servicing the area with infrastructure, as well as the expected building plot sales incomes, are summarised in Table 1.

Excluding the opportunity cost of capital, the estimated costs and incomes indicate a deficit of 3.5 million EUR when all costs are taken into account. If the service infrastructure costs are excluded, the project becomes just about profitable, with a profit of 200,000 EUR. Looking at the cost and income estimates, it can be determined why the private landowner, a subsidiary of the publicly traded forest industry corporation, was not keen to develop the area. The estimated land value increment after the remediation costs would have been 13 million EUR, which equals around 38% of the gross development value (GDV).[7] However, as LUBA makes private landowners responsible for participating in the municipal infrastructure provision costs benefiting them, it is expected that the landowner would have had to contribute to the public infrastructure provision too.

As the developer obligations in Finland are practically always defined in a voluntary agreement between the municipality and the developer, it is impossible to provide a fully accurate estimate of what the landowner would have had to pay. If we assume, following Falkenbach et al. (2021), that the municipality would have been happy to capture 50% of the land value increment after the land remediation costs, the landowner's profit would have been 6.5 million EUR (around 19% of GDV). Although this figure is slightly higher than a typical real estate development profit, 15% of GDV (Ratcliffe et al., 2021), a land development project with an estimated timeframe of 25 years is quite far from a typical real estate development project. Also, the simplified calculation above does not account for the cost of capital

---

[6] These were amendments to the existing local detailed plan as the area already had a legally binding local detailed plan allowing the existing industrial use. Once an area has a legally binding local detailed plan any local detailed plans for such an area are regarded as amendments in the Finnish planning system.

[7] The estimated value of the building plots.

over the 25-year project time. Considering the opportunity cost of capital, the project would most likely have been unattractive for a private profit-seeking actor.[8]

From the municipality's financial perspective, the attractiveness of different approaches depends highly on whether it would take the capital costs related to public land development into account. Had the municipality selected the alternative approach and agreed to capture the assumed 50% of the land value increment after the land remediation costs, the income-cost balance would have been negative 6.3 million EUR after the technical infrastructure provision and negative 10 million EUR after the service infrastructure provision. Thus, to avoid the additional risks related to the public land development approach, the municipality would have had to pay a significant price in terms of lost land value capture. However, the actual capital invested in the project would have been only 16.5 million EUR (including service infrastructure) compared to 37.7 million EUR in the public land development approach. From these figures, it can be derived that the capital costs, which are not accounted for in the simplified calculation above, are significantly higher in the public land development approach.[9] To the authors understanding, it is still relatively common that municipalities do not take these capital costs into account when evaluating their land development projects. The Penttilänranta case demonstrates how very sensitive the financial incentives for public land development are to the assumptions made on the opportunity cost of capital.

3   Penttilänranta Actors

As in any development project employing the public land development approach in Finland, the most important actor of the Penttilänranta case, has been the municipality. In every development project in Finland, the municipality is the planning authority whose approval is needed before any significant development can happen. Before the detailed planning and implementation of the project started, the private landowner was also an important actor. Its continued reluctance to engage in the development of the vacant contaminated site was the catalyst for the municipality to decide to use the public land development approach.

As also typical in the public land development approach, after the municipality had acquired the land, it had effectively three roles in the project: the planning authority, the landowner, and the land (but not building) developer. To implement the project,

---

[8] The calculation of internal rate of return (IRR) would require assumptions on the timing of the cash flows. If assuming land remediation takes 4 years (as in the project plan) and only changing the timing of land sales and related land use fee payments, the IRR for the private developer would vary from 27%, in the case all land sales occurred and fees would be paid at the end of year 4, to 2.4%, in the case the land sales would take place evenly during project years 4–22 as in the project plan.

[9] Again, calculating an IRR requires many assumptions. If following the project plan and assuming land sales to occur evenly over the project years, the IRR would be 0.09% if excluding the costs of service infrastructure and − 1.0% if including the costs of service infrastructure.

the municipality created a separate project organisation with a project manager and a steering group with representation from different departments of the municipality. However, there were no special arrangements for the formal decision-making related to the project. To the authors understanding, creating a specific project with a project manager coordinating different activities related to planning and development, but without authority to deviate from the usual decision-making processes of the line organisation, is relatively typical in Finland in large-scale development projects like this.[10]

The other main actors involved in the project after the land acquisition have been the numerous private building developers who have acquired serviced building plots from the municipality after the completion of the land development. This is the typical role of private developers in projects utilising the public land development approach in Finland. Some private developers to whom a building plot or plots have been reserved already at the planning stage have been involved in the preparation of the local detailed plans. This has been an emerging approach in Finland to arrange the preparation of the local detailed plans in large-scale public land development projects (Valtonen et al., 2017a). In addition to private developers, some building plots have been sold to providers of social/affordable housing.[11]

In addition to the above-mentioned main actors in the projects, there have been various other actors involved in the project, including providers of different technical utilities. As usual in Finland, these utility providers, independent of whether they are owned by the private or public sector, must cover their investments and operating costs by user charges. Thus, although the role of these providers is significant from the perspective of the developed properties having working utilities, the costs or incomes of these actors are not significantly relevant to the municipality from the land policy perspective as they are not part of the public cost recovery or land value capture in Finnish land development.

4 Institutional Perspective

In Finland, the approval of a local detailed plan, as the requirement for any significant development, is a legally binding decision regulating how land can be used. As the landowner was uninterested in changing the use of the site, there was no pressure to amend the local detailed plan from the landowner. After acquiring the site, the municipality had a dual role in the planning process. It was the planning authority preparing and approving the main regulation that affected its opportunities as the

---

[10] Unfortunately, there is no reliable data on how common this form of project organisation is.

[11] There are several formats of social/affordable housing in Finland. Usually, the developer of social/affordable housing can receive subsidies from the state (The Housing Finance and Development Centre of Finland 2024). The most common format of social/affordable housing in Finland is a residential property owned and developed by a public or private registered provider with a lower-than-market interest rate loan from the state. These properties must be kept, for a specified time, in rental use with controlled rents, and the residents have to be selected by social criteria.

landowner to use the land. This is a typical situation in the public land development approach not only in Finland but in every country that has planning control by the local authority that also conducts public land development (Candel & Paulsson, 2023; van der Krabben & Jacobs, 2013; Woestenburg et al., 2019).

Finnish municipalities can generally decide their land policies without any direction from the other levels of the government. There is no legislation or other formal higher-level policies on when the approach can/should be used or not. The use of the public land development approach is promoted as the best practice for the development of greenfield land by the Association of Finnish Local and Regional Authorities (AFLRA), an advocate for municipalities (AFLRA, 2024a). Many Finnish municipalities actively engage in public land development.

4.1 Institutional Advantages of Public Land Development

There are some important institutional factors that incentivise municipalities to use the public land development approach in Finland. These incentives apply, to some extent, to the Penttilänranta case. First, there is an economic incentive related to the public cost recovery. In Finland, the municipality becomes liable for providing public infrastructure in the area after the approval of the local detailed plan, regardless of the ownership. In the public land development approach, the municipality can capture close to 100% of the land value increment to cover these costs.

The high share of the captured value increment is significantly supported by LUBA, which gives municipalities quite an open-ended mandate to use the compulsory purchase for the coordinated development of the municipality. Furthermore, the Act on the Redemption of Immovable Property and Special Rights excludes most of the value increase related to the planning decision from compulsory purchase compensations. When the compulsory purchase legislation is combined with the strong planning position of municipalities provided by LUBA and their strong role in the land markets, the cost of greenfield land for development remains quite modest (Lönnroth et al., 2024). Thus, in the context of greenfield land, the economic incentives for municipalities to use public land development can be significant.

As the calculations above demonstrate, the situation changes in the context of brownfield land, at least after accounting for the opportunity cost of capital. However, the institutional limitations of the alternative options can still make these options better from the risk-profit perspective. In addition to the direct economic incentives, the public land development approach includes several other advantages (Valtonen et al., 2018; van der Krabben & Jacobs, 2013). One of the most important advantages, generally in Finland and also in the Penttilänranta case, is the stronger control of plan implementation. LUBA provides quite limited means for the Finnish municipalities to control the implementation of the private building plots in the local detailed plans if they do not want to pursue this by first acquiring the land. In the next section, we discuss these limitations in more detail.

4.2 Institutional Limitations of the Alternative Approaches

In the Penttilänranta case, the municipality effectively had two alternative policy routes it could have considered had it not followed the public land development approach with voluntary land acquisition: (1) public land development through compulsory acquisition of the land or (2) development of privately owned land with either negotiable or non-negotiable developer obligations.[12]

As mentioned above, Finnish municipalities have quite extensive rights to acquire land by compulsory purchase for the coordinated development of the municipality. Thus, the municipality could have decided to acquire the land by compulsory purchase without the liability for contaminated land. However, in such a case, it is not certain if it could have ultimately avoided the land remediation costs as the pollution of the area had occurred before the Environmental Protection Act came into force in 2000. According to Luntinen (2002), the remediation responsibility is not completely clear in these situations. Even if this approach had been successful in forcing the landowner to cover the costs of land remediation, it is probable that the process would have taken quite a significant time due to the legal complexity and uncertainty.

Generally, when the land of any other landowner than the municipality is developed, LUBA makes the landowner liable to partake in covering the public infrastructure costs. The primary method to facilitate this, according to LUBA, is a negotiable land use agreement (*maankäyttösopimus*). These agreements are considered voluntary agreements that fall within the general contractual freedom in Finland. LUBA explicitly states that these agreements are not limited by the restrictions for the non-negotiable developer obligations in LUBA. Within this context of the highly flexible land use agreements, the usual practice in Finland is to determine the landowner contributions in terms of the estimated value of the building right the landowner receives in the local detailed plan, not in terms of estimated or realised public infrastructure provision costs the plan causes to the municipality (e.g. Falkenbach et al., 2021, 2023).[13] However, municipalities do use rough infrastructure cost estimates to assist with decisions to enter land use agreements, as well. As it is responsible for providing the infrastructure, the municipality usually carries the risk related to the infrastructure provision costs when the land use agreements are used.

As negotiable voluntary contracts, land use agreements only work when the landowner is interested in developing its land, which was not the situation in the Penttilänranta case. Thus, the chances of reaching a land use agreement with terms that would still be competitive against the public land development approach regarding

---

[12] LUBA also allows the municipality to designate a specified area as a special development area with several special arrangements, such as non-negotiable developer obligations that do not have to follow the stringent restrictions that such obligations have to follow in other areas. According to Falkenbach et al. (2021), this land policy instrument has been very rarely used. Their findings suggest that reasons for this include the perceived complexity of the process and the unclarity regarding the incentives for landowners to develop within the procedure.

[13] The share of value increment is usually quite close to 50% (see e.g. Falkenbach et al. 2021, 2023).

the land value capture and cost recovery potential or the municipality's financial risk exposure were not high.

If a land use agreement cannot be reached between the landowner and the municipality, LUBA provides municipalities with a non-negotiable developer obligation that they can impose. This development charge (*kehittämiskorvaus*) is regulated in LUBA with great detail. The most important limitations with the scope of the charge are that (1) it cannot cover service infrastructure costs and (2) it cannot exceed 60 per cent of the estimated land value increment caused by the local detailed plan.

Arguably, the greatest limitation to the applicability of the development charge is that it becomes due only after the landowner has (1) applied and received a building permit or (2) sold the land. This limitation makes the development charge quite a risky land policy instrument as the revenue streams from the development charge are completely conditional on the future decisions of the landowner. This would have been the situation in the Penttilänranta case as well had the municipality decided to go through this approach. If the landowner had not developed the site two years after the approval of the local detailed plan, LUBA would have allowed the municipality to impose a reminder to build (*rakentamiskehotus*) on the building plots in the plan. If the landowner had not developed a building plot subject to the reminder to build three years after receiving the reminder, the municipality would have automatically received the right to use compulsory purchase for the acquisition of the building plot. Had it decided to use compulsory purchase, the municipality would have become the landowner, but many years later than in the public land development approach. As mentioned above, it still could have even ended up paying the land remediation costs, due to the timing of the polluting activity.

It can be concluded that the institutional framework, with a strong reliance on municipal land ownership, placed the municipality in a position where it was challenging to facilitate the development by any other means than the public land development approach. It most probably was the municipality's only effective option. It is unclear if there could have been a route to realise the development without the municipality taking the land remediation costs. However, it is quite clear that the alternative routes would have made the development process longer, more complicated and more uncertain as it would have involved dealing with a landowner reluctant to develop. There could have also been lengthy litigation over the contamination remediation liabilities between the parties.

5   Reflection on the Land Policy Debate

The discussed case is well representative of the land policy in Finland. The public land development approach is widely used, traditionally in the greenfield development context but nowadays increasingly in brownfield redevelopment, too (Krigsholm et al., 2022; Lönnroth et al., 2024; Valtonen et al., 2017a; Viitanen et al., 2003). The Penttilänranta case is one of the projects where the approach has been used in this emerging context. What makes this case atypical is that the municipality did not own, directly or indirectly, the brownfield site in question. Instead, the municipality bought the site and all its environmental risks. There are many brownfield areas in Finland where the public land development approach has been used out of necessity, as the municipality already was the ultimate owner or polluter of the site. For example, in the capital city of Helsinki, there are several former industrial harbour areas that have been redeveloped. However, these areas were mostly owned and occupied either by the municipality or municipal agencies or companies, making the municipality eventually liable for the environmental remediation (City of Helsinki, 2008a; 2008b).

In a broader policy debate perspective, outside the academia, the discussion of the extent to which the public land development approach should be employed or the financial risks the municipalities are exposed to while employing it has been mild at best. One strong contributor to the absence of such debate is undoubtedly the AFLRA which quite strongly recommends using public land development to secure the effectiveness of land use planning, building plot supply, and financial viability of infrastructure provision (AFLRA, 2024a). Thus, the motives and practices of conducting public land development often largely avoid scrutiny in the Finnish policy discussion.

The Penttilänranta case is not controversial from the property rights and market intervention perspective as both parties were willing to act voluntarily. The private landowner wanted to sell the site and related environmental liabilities, and the municipality wanted to acquire them. However, the use of the public land development approach in this case highlights the question that has been debated elsewhere but not that much in Finland; to what extent should public money be used to implement these kinds of risky redevelopment projects, especially when it involves taking the land remediation liabilities from a private business that has polluted the land? For example, in the Netherlands, there has been a notable shift towards private land development after the financial troubles many municipalities faced with their public land development projects in the aftermath of the 2007–2008 Global Financial Crisis (van Oosten et al., 2018; Woestenburg et al., 2018). Such questioning of public land development has not really emerged in Finland. This is most probably related to the compulsory purchase legislation and the strong planning position of the municipalities keeping the land acquisition prices of greenfield land relatively low, even in the areas of the highest growth (Lönnroth et al., 2024). Furthermore, the quite highly decentralised fiscal system of Finland[14] probably has a role to play in stifling the debate and discussion of the financial risks involved in the public land development approach as the municipalities are less reliant on public land development revenues or state grants (Götze & Hartmann, 2021; Valtonen et al., 2017b). As the Penttilänranta case demonstrates, these questions become much more difficult to avoid when the public land development approach is used in the brownfield redevelopment context instead of greenfield development.

---

[14] According to AFLRA (2024b), on average, around one half of the municipal revenues come from the local taxes, including municipal income tax, corporate tax and property tax.

There is a fine balance between the acceptable financial risk of urban redevelopment and the need to facilitate development when private market actors are uninterested in developing the area. As the Penttilänranta case demonstrates, in the Finnish institutional framework, it can be challenging, if not impossible, for a municipality to find any other way than the public land development approach to facilitate development in these situations. As the whole planning and development system is so inherently built around the idea of the public land development approach, many alternative approaches, such as public–private joint ventures that are common elsewhere (de Paula et al., 2023), are very rare in Finland. It is interesting to see if this will change in future when brownfield redevelopment will presumably become a more and more common context for Finnish land development. It will also be interesting to see whether some new land policy instruments, such as compulsory land readjustment (Gozalvo Zamorano & Muñoz Gielen, 2017; Muñoz Gielen & Mualam, 2019), will emerge, allowing municipalities to facilitate development more effectively when the land is owned by a landowner reluctant to develop.

To conclude, the Penttilänranta case highlights how Finnish municipalities have very limited alternatives for the public land development approach when dealing with land use problems resulting from the reluctance of the landowner to develop. As these problematic situations tend to take place in brownfield contexts with high up-front land remediation costs, this exposes the municipalities to significant financial risks. Due to the environmental pressures to favour brownfield redevelopment over greenfield development, Finnish municipalities will presumably find themselves increasingly in situations similar to the Penttilänranta case. If the institutional framework remains as it is, the public land development approach, which arguably has served Finnish municipalities well in greenfield development, will stand strong in the brownfield redevelopment context, too.

References

Act on the Redemption of Immovable Property and Special Rights. 603/1977. The Parliament of Finland.

Association of Finnish Local and Regional Authorities (AFLRA). (2024a). Maapolitiikan opas. Retrieved July 16, 2024, from https://www.kuntaliitto.fi/julkaisut/maapolitiikan-opas

Association of Finnish Local and Regional Authorities (AFLRA). (2024b). Verotus. Retrieved July 16, 2024, from https://www.kuntaliitto.fi/talous-ja-elinvoima/verotus

Candel, M., & Paulsson, J. (2023). Enhancing public value with co-creation in public land development: The role of municipalities. *Land Use Policy, 132*, 106764. https://doi.org/10.1016/j.landusepol.2023.106764

City of Helsinki. (2008a). Sörnäistenrannan ja Hermanninrannan osayleiskaava - Osayleiskaavan selostus. City of Helsinki.

City of Helsinki. (2008b). Jätkäsaari, osayleiskaava – selostus. City of Helsinki.

City of Joensuu (2008). Penttilänrannan hankesuunnitelma. City of Joensuu

de Paula, P. V., Marques, R. C., & Gonçalves, J. M. (2023). Public–private partnerships in urban regeneration projects: a review. *Journal of Urban Planning and Development, 149*(1). https://doi.org/10.1061/JUPDDM.UPENG-4144

Environmental Protection Act. 86/(2000). The Parliament of Finland.

Falkenbach, H., Krigsholm, P., Lönnroth, T., Puustinen, T., Ekroos, A., Häkkänen, M., Kortelainen, J., Luomala, N., Rausmaa, S., & Toivanen, M. (2021). Maapolitiikan nykytila ja tulevaisuus: Keinot, vaikuttavuus, avoimuus ja hyväksyttävyys. Prime Minister's Office.

Falkenbach, H., Krigsholm, P., Kuivalainen, H., Firoozi Fooladi, B., & Pyykkönen, S. (2023). Kaavoituksen vaikutukset kiinteistöjen arvoihin ja maankäyttöpoliittisten keinojen muutosten vaikutusarviointi. Prime Minister's Office.

Gozalvo Zamorano, M. J., & Muñoz Gielen, D. (2017). Non-negotiable developer obligations in the Spanish Land readjustment: An effective passive governance approach that 'de facto' taxes development value? *Planning Practice & Research, 32*(3), 274–296. https://doi.org/10.1080/02697459.2017.1374669

Götze, V., & Hartmann, T. (2021). Why municipalities grow: The influence of fiscal incentives on municipal land policies in Germany and the Netherlands. *Land Use Policy, 109*, 105681. https://doi.org/10.1016/j.landusepol.2021.105681

Muñoz Gielen, D., & Mualam, N. (2019). A framework for analyzing the effectiveness and efficiency of land readjustment regulations: Comparison of Germany. *Spain and Israel. Land Use Policy, 87*, 104077. https://doi.org/10.1016/j.landusepol.2019.104077

Krigsholm, P., Puustinen, T., & Falkenbach, H. (2022). Understanding variation in municipal land policy strategies: An empirical typology. *Cities, 126*, 103710. https://doi.org/10.1016/j.cities.2022.103710

Land Use and Building Act. (LUBA). 132/1999. The Parliament of Finland.

Luntinen, M. (2002). Kunta ja pilaantunut maaperä. Association of Finnish Local and Regional Authorities.

Lönnroth, T., Krigsholm, P., Falkenbach, H., & Oikarinen, E. (2024). Advancing understanding of the linkages between local land policy interventions and the responsiveness of housing supply: Intervention mechanisms in the Finnish context. *Land Use Policy, 141*, 107157. https://doi.org/10.1016/j.landusepol.2024.107157

Ratcliffe, J., Stubbs, M., & Keeping, M. (2021). *Urban planning and real estate development.* Routledge.

Statistics Finland. (2024). Dwellings and housing conditions. Table 116a - Household-dwelling units by number of persons and type of building, 1985–2023 . Statistics Finland. https://stat.fi/en/statistics/asas

The Housing Finance and Development Centre of Finland. (2024). Rahoitus uudisrakentamiseen. Retrieved July 17, 2024, from https://www.ara.fi/fi/yhteisot-ja-yhtiot/rahoitus-uudisrakentamiseen

Valtonen, E., Falkenbach, H., & Viitanen, K. (2017a). Development-led planning practices in a plan-led planning system: Empirical evidence from Finland. *European Planning Studies, 25*(6), 1053–1075. https://doi.org/10.1080/09654313.2017.1301885

Valtonen, E., Falkenbach, H., & Van der Krabben, E. (2017b). Risk management in public land development projects: Comparative case study in Finland, and the Netherlands. *Land Use Policy, 62*, 246–257. https://doi.org/10.1016/j.landusepol.2016.12.016

Valtonen, E., Falkenbach, H., & Viitanen, K. (2018). Securing public objectives in large-scale urban development: Comparison of public and private land development. *Land Use Policy, 78*, 481–492. https://doi.org/10.1016/j.landusepol.2018.07.023

Viitanen, K., Palmu, J., Kasso, M., Hakkarainen, E., & Falkenbach, H. (2003). *Real estate in Finland.* Helsinki University of Technology.

Van der Krabben, E., & Jacobs, H. M. (2013). Public land development as a strategic tool for redevelopment: Reflections on the Dutch experience. *Land Use Policy, 30*(1), 774–783. https://doi.org/10.1016/j.landusepol.2012.06.002

Van Oosten, T., Witte, P., & Hartmann, T. (2018). Active land policy in small municipalities in the Netherlands: "We don't do it, unless..." *Land Use Policy, 77*, 829–836. https://doi.org/10.1016/j.landusepol.2017.10.029

Woestenburg, A. K., van der Krabben, E., & Spit, T. J. (2018). Land policy discretion in times of economic downturn: How local authorities adapt to a new reality. *Land Use Policy, 77*, 801–810. https://doi.org/10.1016/j.landusepol.2017.02.020

Woestenburg, A., van Der Krabben, E., & Spit, T. (2019). Legitimacy dilemmas in direct government intervention: The case of public land development, an example from the Netherlands. *Land, 8*(7), 110. https://doi.org/10.3390/land8070110

Eero Valtonen is a Lecturer in Real Estate at the Department of Planning, Property and Environmental Management of The University of Manchester. At The University of Manchester, he is also a member of The Manchester Real Estate and Urban Economics (MREUE) research group. Eero holds a Ph.D. in Real Estate Economics from Aalto University, and his main research interests focus on public land policies, especially their connections to housing markets. His work has been published in esteemed academic journals, including European Planning Studies, Land Use Policy, Regional Science and Urban Economics and Town Planning Review.

Heidi Falkenbach is an Associate Professor in Real Estate Economics at the Department of Built Environment of Aalto University. Heidi's main research interests include the economics of real estate markets, real estate finance, and effects of public policies on real estate markets. Her work has been published in esteemed academic journals, including Cities, International Journal of Urban and Regional Research, Land Use Policy, European Planning Studies, Journal of Real Estate Finance and Economics and Journal of Portfolio Management. Along her academic research, she also actively engages in policy advisory and evaluation.

Land Policy in France: Tensions Between National Policies and Local Practice

Camille Le Bivic, Sonia Guelton, Mathias Jehling, and Caspar Kleiner

Abstract The case of Bouleurs, a suburban municipality in the Paris region, provides insights into the institutional context of land policies in France. It highlights the tensions between national and local authorities when applying urban development instruments, especially within the context of no-net-land take goals. Through this case, the interaction between national land policy goals and the range of instruments used at the local level are examined. The implementation of these policies in a small municipality underscores the importance of the local context, while also revealing broader challenges in planning and managing land development. Tensions arise between the relatively high degree of flexibility afforded to local authorities in applying land policy instruments and the overarching national land policy goals as well as between the diverging interests of public actors and private landowners.

1 A French Understanding of "Land Policy"

Land policy in France emerged in the 1960s in response to rural-to-urban migration, which led to the fragmentation of agricultural properties and the expansion of settlements in the urban periphery (Fig. 2). The national government sought to balance the need to secure agricultural production with the need to coordinate spontaneous urban expansion. This led to the creation of specialised agencies, such as the Land Development and Rural Development Agency, SAFER (*Société d'Aménagement Foncier et d'Établissement Rural*) through the Agricultural Orientation Law of 1960 (*Loi d'Orientation Agricole, 1960*). SAFER's primary role is to control the agricultural

C. Le Bivic
Fédération Nationale des Safer, Paris, France
e-mail: camille.lebivic@safer.fr

S. Guelton (✉)
Université Paris Est Creteil, LAB'URBA, F-94010 Créteil, France
e-mail: guelton@u-pec.fr

M. Jehling · C. Kleiner
Leibniz Institute of Ecological and Regional Development, Dresden, Germany
e-mail: m.jehling@ioer.de

© The Author(s) 2025
T. Hartmann et al. (eds.), *Land Policies in Europe*,
https://doi.org/10.1007/978-3-031-83725-8_7

land market in rural areas and to prevent the conversion of agricultural land by using instruments such as land purchase and property consolidation (Code Rural et de La Pêche Maritime, 2020; Shields, 2023) (Fig. 2).

While SAFER operates in agricultural areas to manage land use, no equivalent agency exists to control the land market in urban or urbanising areas. The Land Orientation Law of 1967 (*Loi d'Orientation Foncière*, 1967) began to structure French territorial development by introducing key instruments, many of which remain relevant today despite major readjustments in the early 2000s. The law continues to provide the foundational framework for planning instruments that enable public actors to develop and manage urban development projects in collaboration with private stakeholders.

The Land Orientation Law introduced two key elements: development plans and property focussed instruments. First, development plans outline territorial strategies and provide the framework for development pathways. At the regional level

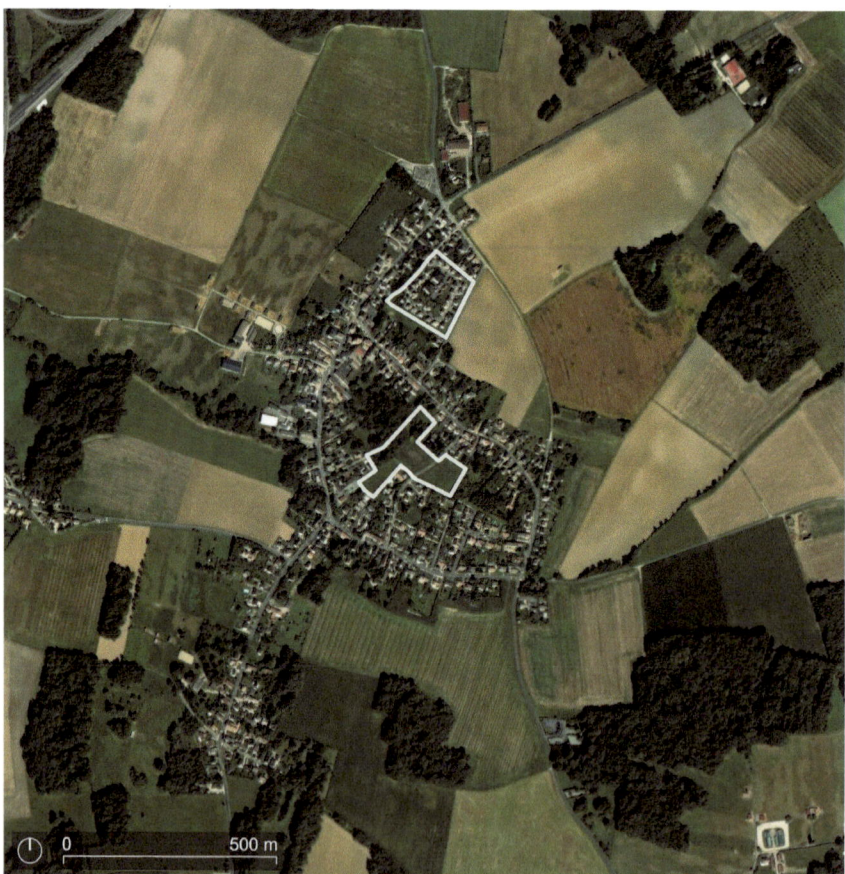

Fig. 1 Bouleurs in the Greater Paris Area, France (map: Kleiner, Jehling 2024, Aerial Imagery Provider © Esri, Maxar, Microsoft)

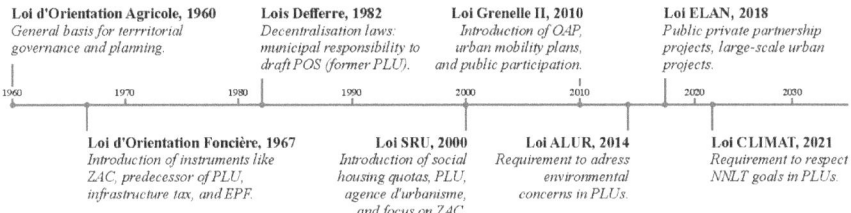

Fig. 2 Evolution of land policies in France

nowadays, Territorial Cohesion Schemes, SCOT (*Schéma de Cohérence Territoriale*), guide land use planning by addressing key aspects such as mobility, the environment, economic activities, and housing in a comprehensive document. At the municipal level, Land Use Plans, PLU (*Plan Local d'Urbanisme*), reflect the SCOT and set binding land use rules, including specifications for building types, height, density, and architectural requirements. These plans can also be developed collaboratively between municipalities through Inter-municipal Land Use Plans PLUi (*inter-communal*). Secondly, a range of property-focussed instruments are used to enact these territorial strategies. They range from taxes, such as property and development tax, to land development contracts between public and private stakeholders, such as those used for Joint Development Zones, ZAC (*Zone d'Aménagement Concerté*).

In France, this legislative context is commonly understood as land policy, characterized by a set of planning and property-oriented instruments designed to achieve politically defined goals (Comby & Renard, 1990). Recent legislative changes have introduced new instruments focusing on emerging political objectives, such as the no-net-land-take (NNLT) objective introduced in 2021, which requires local authorities to control land take with quantitative targets. Instead of a single, coherent strategy for implementing policy goals in France, instruments are applied based on the specific governmental level, its interests, and its degree of autonomy. This fragmented approach often leads to conflicting objectives, particularly concerning private property rights and development pathways. In Bouleurs, for instance, a conflict arises between the national NNLT goals and the strategic application of instruments to attract new residents.

Historically, the national government was the primary public actor in defining territorial development strategies and applying land policy instruments. However, a lack of coordination among various ministries often led to conflicting strategies, as the ministries for housing, economic development, infrastructure, or ecology, issue their own land policy strategies and guidelines without considering possible contradictions (Figeat, 2016). The decentralisation process initiated by the Defferre laws (*Lois Defferre*) in the 1980s increased the power of municipalities, granting them full responsibility for urban development, including the development of regional and inter-municipal strategies and the implementation of land policy instruments. However, decentralisation has not eliminated the need for coordination among municipal strategies, as comprehensive inter-municipal strategies remained weak. In response, inter-municipal public structures have been granted several powers that are "shared" among municipalities. For instance, since 2021, many municipalities have been required to collaborate on PLUi. Nevertheless, the responsibility for assessing and granting building permits remains the responsibility of each individual municipality. Also, for land policy within municipalities it is challenging to identify a consistent trend across France. Surveys (e.g. Le Bivic & Melot, 2020) suggest that several municipalities implement land instruments, property taxes, or public development projects, without strategically coordinating them.

While the decentralization process has empowered municipalities, the national government retains significant influence. Through various instruments, the central government can intervene in local decisions when 'national interests' are at stake, particularly in controlling land take or preserving natural land. Recent laws (Fig. 2), such as the Law for Access to Housing and a Renewed Urbanism, Loi ALUR (*Loi Pour l'Accès au Logement et un Urbanisme Rénové*, 2014), the Law on the Progression of Housing, Development and Digital Technology, Loi ELAN (*Loi Portant Évolution du Logement, de l'Aménagement et du Numérique*, 2018), or the more recent Climate and Resilience Law (*Loi CLIMAT*, 2021) which introduced the NNLT objective ZAN (*Zero Artificialisation Nette*) (Guelton, 2018), have further constrained local powers by introducing compulsory objectives for local decision-making.

As a result, land policy in France offers public actors a wide array of instruments, but these instruments lack a single, coherent purpose. Various levels of government, from local to national, develop and implement their own land policy strategies, risking to result in case-by-case coordination, particularly in larger municipalities or key political areas. This can be seen in the Paris region, where the national government is highly influential, or in large metropoles which have competent administrations with a high level of strategic awareness. In most cases, however, land policy is subject to local interpretation and strategic application. To understand the forces shaping the implementation of land policies at the municipal level, we examine a case within the periphery of the Paris region, where diverging land policy goals—such as housing production and land preservation—present significant challenges in the application of policy instruments. In France, the periphery of agglomerations, also referred to as suburban or peri-urban areas, is home to 45% of the French population and encompasses two-thirds of the municipalities in France (Couleaud et al., 2021).

2 Local Implementation of Land Policy: The Case of Bouleurs

The focus of this case is Bouleurs, a small municipality in the eastern suburb of the Greater Paris metropolitan region. Bouleurs (see Fig. 1) is one of around 1000 small municipalities on the outskirts of Paris, making up 74% of the total of municipalities of the capital region. It is important to note that French municipalities are generally small: in 2017, of the 34,968 total municipalities, 72% had fewer than

1000 inhabitants and 25% had fewer than 200 inhabitants. Despite their size, each municipality has a mayor and council responsible for a wide range of duties, particularly in implementing land policy and urban development.

Among the peripheral municipalities of the Paris region, three types can be distinguished based on their proximity to Paris. First, those close to the core of the metropole have reached a high degree of urbanisation, leaving limited opportunities for further expansion. In these areas, housing prices are high, and urban renewal projects focused on densification are predominant. Second, there are municipalities about 100 km from Paris that are less attractive due to their poor public transport access. Here some mid-sized cities, often former industrial centres or administrative or commercial hubs, face a degree of decline as they struggle to compete with central areas, closer to Paris. Their prevalent strategy is to "fill the empty plots". Finally, municipalities like Bouleurs represent the third category of Parisian suburbs. These areas have extensive reserves in agricultural, natural, and preserved land but are also close enough to Paris to attract significant development pressure. Ease of access to Parisian industries and services generates high demand for housing and other activities, leading to urban sprawl. This sprawl brings negative effects including transport-related carbon emissions and environmental degradation, as well as greater demand for public services and infrastructure. National land policies specifically target these issues, focusing on commuting, biodiversity loss, the consumption of agricultural land, and landscape degradation (Le Bivic & Melot, 2020). Urban sprawl is also seen in several other French metropoles and mirrors a trend observed across most European countries (Guelton et al., 2011).

Bouleurs is part of the inter-municipal cooperation known as the *communauté d'agglomération Coulommiers, Pays de Brie*. A rural municipality with about 1700 inhabitants in 2020, it has a population density of 206 inhabitants/km^2, compared to the average of 479 inhabitants/km^2 in the entire peripheral area. Due to its attractiveness within the capital region and a shortage of housing, Bouleurs has experiences rapid population growth (2.3% per year between 2014 and 2020) and an increase in dwellings (2.6% per year). In comparison, population growth in the entire capital region was 0.3% per year, with a 0.95% annual increase in dwellings during the same period (Rannou-Heim, 2022). This growth is largely driven by good urban transport connections, including to the employment hub of Marne-la-Vallée. Housing prices in Bouleurs reflect its attractiveness: the mean housing price is 3000 €/m^2, or about 200,000 € for a detached house, while prices in Marne-la-Vallée are 25% higher, and significantly higher in the closer Parisian suburbs (Hilal et al., 2018). With a 45 km distance to Paris and a 15 km distance to Disneyland Paris in Marne-la-Vallée, Bouleurs is appealing to commuters. Despite its small size, Bouleurs exemplifies the phenomenon of urban sprawl. Like many municipalities in this historically agricultural, peripheral area, Bouleurs is transforming under demographic pressure. Many households from Paris or Marne-la-Vallée are moving to Bouleurs in search for a rural lifestyle and for larger, more affordable dwellings.

The municipality is currently facing two potentially conflicting urban development objectives: on the one hand, it aims to produce affordable housing and attract new residents—partly to meet the municipal goal of keeping the local school open. On the other hand, it seeks to preserve natural, forested, and agricultural land, aligning with national policy goals like the NNLT objective. Here, challenges arise from spontaneous, privately driven urban development projects that often clash with the municipal land policy strategy. This leads to the consumption of natural land and inefficient allocation of public investments in infrastructure and services. Until the 2010s, a consistent land policy strategy was not a priority for municipal decision-makers. However, with the national push for NNLT goals in 2021, the need to develop a strategy has become more pressing.

The urban development project Rue Romain Rolland in Bouleurs is a typical example of a private development project on a greenfield at the urban fringe. Based on empirical material (Le Bivic, 2018), this case illustrates how land policy instruments have been applied over the past decade. For the project Rue Romain Rolland, a developer secured agreements with several landowners willing to sell their land. However, the municipality's local urban plan (PLU) had already identified another zone in the town centre for infill development (zone further south on the map in Fig. 1). When the municipality found the prices demanded by the majority of landowners in this central zone too high, they shifted focus to a peripheral area (zone north on the map). Although this second area, Rue Romain Rolland, was less favourable due to its less accessible location, it was nevertheless planned for urban development with single family homes and apartments (Fig. 3). Originally, this area was intended as a long-term development zone, requiring another revision of the PLU before development could proceed. However, it was chosen because it was easier to develop, with fewer landowners involved and promises of sale more readily obtained by the developer.

The project highlights the challenges arising from the lack of active land policy strategy which cannot be reduced to the efficiency or effectivity of planning instruments. The interaction between public actors and property owners is central to understanding these challenges. The case of Bouleurs offers valuable insights into how elected officials, under pressure from urban developers and landowners, may abandon their strategy of infill development to instead rely solely on the land policy instruments available to them to implement private initiatives. It also reveals that the municipal public actors face limited institutional capacity and resources.

Fig. 3 Bouleurs, Rue Romain Rolland (*Photo* Camille Le Bivic)

3   Actors Involved in Bouleurs, Rue Romain Rolland

3.1   Public Actors

For the project, the municipality remains the main public actor. In rural Bouleurs, it often welcomes any kind of development along the main road and on greenfields. Although the mayor is eager to concentrate new constructions in the municipal centre, there is significant interest in developing agricultural land (Le Bivic, 2018). Urban developments, while generating tax revenue, increase expenses due to the need for infrastructure investments. New residents require public services such as roads, sewage, drinking water, energy, telecommunications, and waste management. Consequently, land policy aimed at limiting land consumption becomes more strategic, as spontaneous private projects are not curtailed. The municipality's role is more reactive than anticipatory, which can lead to challenges, especially concerning the public budget. Nonetheless, municipalities are working to strengthen their land policy strategies and utilise instruments to negotiate with other actors.

The national government provides the legal framework for municipal planning authorities and sets societal policy goals such as NNLT and environmental protection. However, its involvement in Bouleurs remains implicit, and it is unclear whether the responsibility for achieving these goals rests with the national government or the municipality. At the time of this study, the French Région and Département levels lacked competencies in spatial planning. Since 2021, the regional level has acquired new responsibilities and resources to translate NNLT goals at the sub-national level.

3.2   Private Actors

At Rue Romain Rolland, current landowners control access to the land necessary for new development. Several factors contribute to their power: the municipality's limited financial resources, the close ties between elected officials and landowners due to the small size of the municipality, and the officials' desire to address housing needs quickly. Landowners are willing to sell to developers, aiming to profit from increases in land value resulting from changes in the PLUi. Agricultural land may be sold at €0.5/m², while buildable land can be sold at €150/m² or more. Landowners may not anticipate delays in obtaining building permits or fluctuations in land prices and planning decisions. Despite this, they can still influence the planning process to protect their interests. In Bouleurs, the greenfield development zone allowed the municipality to bypass the high land prices requested by the owners of the infill development zone, enabling it to meet housing needs with affordable prices. As of 2025, the land use in the first identified infill site remains agricultural.

Consultancy offices assist the municipality with the legal aspects of the PLU. In Bouleurs, however, the role of external consultants was minimal as the elected officials already had a clear vision for new urban development.

The project developer is a national private firm specialising in rural areas. This developer has completed the site with approximately 30 detached houses on 500 m^2 plots and two apartment buildings: one for private ownership and the other managed by a social housing agency. The developer focuses on large projects outlined in PLUs and seeks projects in stable real estate markets with sufficient housing demand (Le Bivic, 2018). The developer negotiates with the municipality, looking for flexible land policy instruments such as an OAP (*Orientations d'Aménagement et de Programmation* or Development and Programming Guidelines), which indicates that the plots are of strategic importance and that the municipality is prepared to act. The developer was also informally involved in the revising of the PLUi and zoning rules for greenfield development. Negotiating with the developer requires the municipality to skilfully use policy instruments and to defend its interest. For instance, the municipality negotiated with the developer to meet its strategic goals regarding density.

The interaction among different actors reveals diverging interests and goals. These might lead to conflicts, which can occur not only between public and private actors, but also among different public authorities, e.g. supralocal and local, and private actors, e.g. landowners for or against new development. Negotiation between private and public actors is crucial in the French land planning system. The frequent revision of PLUs creates room for such negotiations.

4   Institutions of Land Policy in France

In the institutional framework of French land policy (Fig. 4), two primary approaches exist for managing development projects like Rue Romain Rolland. The first is a directive, top-down process facilitated by a contract between the public authority and the private developer. This method involves the use of a Joint Development Zone, ZAC (*Zone d'Aménagement Concerté*), which allows for additional measures such as expropriation or reductions in development taxes. The ZAC approach is typically reserved for complex developments. The second option is a more flexible regulatory process based on planning rules and taxes (Booth, 1998). In this case, initiatives are driven by individual promotors or builders, with the primary municipal power being the issuance of building permits (Fig. 4).

For Bouleurs, a ZAC would have provided a clear framework for collaboration between public and private actors. However, since the development only involved private housing—either for purchase or social housing—and lacked public facilities, mixed uses, or long-term development needs, it was not deemed complex enough to warrant a ZAC. Consequently Bouleurs, like many municipalities, preferred managing the development through permits and subdivisions. The ZAC approach requires significant financial and human resources, which many rural municipalities lack (Pichon, 2022).

In addition, elected representatives are increasingly using more flexible instruments ahead of an operational phase of a project. These include identifying urbanisation areas and rules in a PLU combined with an urban development tax or drafting

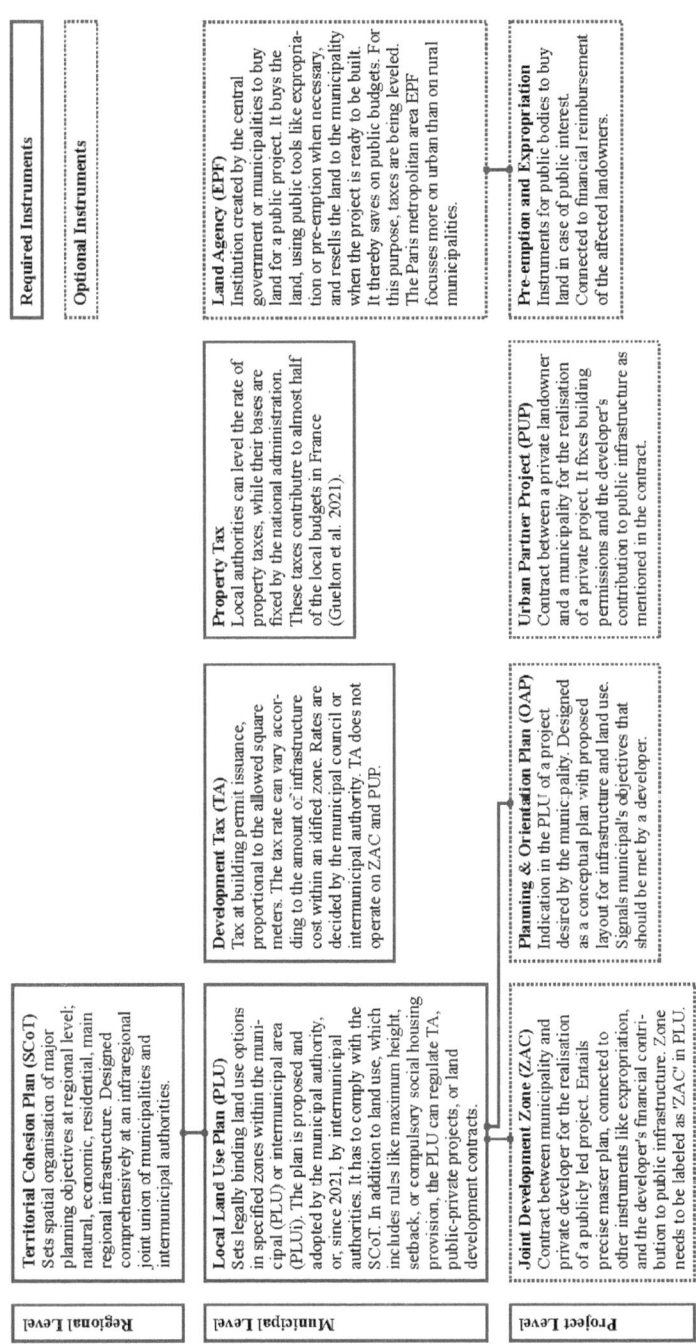

Fig. 4 Institutional framework: Main instruments available to French municipalities (authors' visualization)

of an OAP. These instruments provide enough flexibility to negotiate development details with private actors (see Guelton & Pouillaude, 2023) when they apply for building permits. The inherent vagueness of these regulations compels developers to approach municipal representatives to clarify project details and secure planning permits. This adaptability allows local officials to pursue their spatial plans despite limited resources. However, without active involvement from local officials in negotiations, there is a risk that projects may develop in ways that are undesirable in terms of location, urban form, or alignment with the NNLT objective.

4.1 PLU as Passive Land Policy

The PLU (*Plan Local d'Urbanisme*) is a crucial planning document established by the municipality and plays a central role in the French land policy system. It divides the municipal area into zones designated for various land uses, as defined by the national building code (*Code de l'Urbanisme*). This division addresses current and future urban developments, land uses, and urban forms (Jehling & Hecht, 2022). The PLU sets the planning rules, zoning regulations, and development constraints for these areas. Inter-municipal plans, known as a PLUi, enable cross-municipal planning for coherent development between multiple municipalities. The PLU is legally binding for public and private property owners and serves as the legal basis for granting building permits (Jehling & Hecht, 2022). It can also designate some public uses as private areas, which becomes compulsory if the landowner wants to sell the land. The municipality can amend the PLU without compensating for any decrease in land value (Lacoere et al., 2023). The PLU provides a strong foundation for negotiations between the municipality and as it is legally binding, yet elected, officials have the authority to modify it (Le Bivic and Melot 2020b). Therefore, municipalities can adopt a passive role in development projects, focusing primarily on granting or rejecting building permits. Additionally, municipalities can integrate various tools into the PLU, such as an OAP or initiating a ZAC,[1] to balance public and private interests, guide developers towards public goals, and ensure contributions to public costs (Acosta & Renard, 2018).

4.2 OAP as Reactive Land Policy

An OAP (*Orientations d'Aménagement et de Programmation*) can be introduced by the municipality within the PLU and outlines the municipality's goals and strategies for development, focusing on specific spatial or thematic issues. The instrument is particularly useful in rural municipalities influenced by metropolitan areas,

---

[1] Or an Urban Partner Programme, PUP (*Projet Urbain Partenarial*) in case of a private project on private land.

like Bouleurs. Introduced in 2010 as part of the Grenelle II Law, the OAP allows municipalities to address issues such as housing or transportation. It defines necessary actions and operations by private actors to meet environmental, heritage, or social concerns, specifying road characteristics or specific building morphology and functions. Unique to the OAP is its focus on strategic areas for development, with qualitative elements that are open to interpretation. Although a legally binding document, the OAP's flexibility can create uncertainty regarding construction capacity (number of dwelling units, infrastructure equipment, etc.) and land values in rural municipalities. Some public and private actors argue that regulatory documents like the OAP can hinder housing production and offer limited guidance on qualitative land development that integrates landscape criteria and morphologies (Le Bivic, 2018). Nonetheless, the OAP provides especially smaller municipalities with a low-cost means to address social and environmental concerns.

4.3   ZAC as Active Land Policy

A ZAC (*Zone d'Aménagement Concerté*) represents a specific form of public–private partnership for project development. Its main purpose is to develop an area in line with municipal or governmental policy goals and then sell or lease the land to public or private users. The public actor or another actor by concession manages the development process, including planning studies, construction, land acquisition, land readjustment and sale (*lotification*), and provision of public infrastructure (Jehling & Hecht, 2022). A ZAC can be planned as part of the PLU or initiated independently. Typically, it is initiated by the municipality in collaboration with a private developer, though it may also be initiated by other governmental agencies or private developers. Municipalities often adopt an active land policy, by buying, preempting, or, in crucial situations, expropriating land. While public land ownership can be advantageous in France, many municipalities lack the resources to acquire land due to financial constraints. Expropriation, though potentially useful, involves significant costs and is less frequently employed. In Bouleurs, for example, the process is time consuming and financially burdensome. Land Agencies like EPF (*Établissement Public Foncier*) can assist municipalities in acquiring land, but their focus on small urban projects and public objectives such as social housing may not align with Bouleurs' needs. The ZAC instrument is generally reserved for large-scale projects with significant political importance, allowing for early and collaborative development.

Various regulations and instruments affect land policy. Some, like expropriations, have substantial impacts on private property rights and carry significant risks for governments, leading to their infrequent use. Others, like the PLU, set constraints and obligations for private developers less forcefully. The OAP advises private developers to pursue political goals, while zoning regulations impose mandatory obligations. Although the ZAC is a valuable tool for stabilising land prices and to advancing public projects, its complexity and cost often deter rural municipalities from utilizing it.

Consequently, rural municipalities typically rely on softer influence and ownership-based tools (Guelton & de Flore, 2020).

5 Reflection: Conflicts in Applying Local Land Use Plans Within the National Land Policy Framework

The passive, reactive, and active land policy instruments are closely interlinked institutionally, but not without tensions. The development and revision of the PLU, the key instrument for land management, underscore conflicts among national and local authorities, elected officials, landowners, and urban developers. The case of Rue Romain Rolland in Bouleurs exemplifies the key challenges in land policy in France. These can be structured into four main conflicts related to (a) the degree of flexibility, (b) the local application of land policy instruments, (c) the tension between private and public interests, and (d) the evolving national NNLT policy:

(a) Degree of Flexibility: The choice of the development zones within Bouleurs' PLU shows how instruments defined by the national government, while binding, allow municipalities significant flexibility in implementation. Municipalities can easily adapt these instruments to align with their strategies and objectives, often on a case-by-case basis. However, this flexibility carries the risk that municipalities may not be able to implement land policy goals that align with national or private interests.

(b) Application of land policy by law or by local practice: The case reveals a symptomatic attitude within the public sector. While the central government provides a range of formal instruments, municipalities often struggle to implement land policy effectively. They face constraints in technical capacity, time, and financial resources. As a result, municipalities evaluate these instruments based on their feasibility and the likelihood of community acceptance, leading to a bottom-up approach where land policy is shaped by what is locally agreeable rather than by overarching national objectives.

(c) Land property interest versus public policy: Since the Revolution of 1789, France has upheld a strong tradition of powerful property rights. Courts generally protect full and complete ownership, even as various legislations impose limits on these property rights in the public interest. This tension is particularly evident in land policy, where regulations aim to preserve neighbourhoods, safeguard natural environments, or protect against environmental hazards. Consequently, most mayors are then inclined to regulate developments and implement strong regulations to achieve their objectives.

However, rural landowners, who are often deeply attached to their land as it represents both their livelihood and legacy, resist public interventions. Even when they are no longer farmers, they continue to strongly protect their property rights, often seeking to either pass the land on through family inheritance or benefit from intense urbanisation as their agricultural revenues decline. In small municipalities like Bouleurs, landowners maintain close ties with the municipality, functioning as an effective lobby or a persistent counterbalance to be taken into consideration. Given this dynamic, municipalities are reluctant to impose restrictive land policy instruments on landowners directly. Instead, they negotiate to reconciliate conflicting objectives. The OAP represents an opportunity to shift from strict regulatory control to a more negotiable framework that still upholds public objectives. While the OAP is legally binding, it remains sufficiently vague to allow room for interpretation and adaptation by the involved parties.

France's extensive legislation, designed to preserve individual freedom and interests, sets the institutional framework for land policy, exemplifying the balance between law and individual liberty. The French system grants significant freedom to landowners, and this chapter illustrates the compromises necessary for applying land policy at the local level. Historically, conflicts between public interests and the individual landowners' rights have been limited to specific cases, such as environmental protection or infrastructure developments.

(d) Evolving NNLT policy: Three years after greenfield development at Rue Romain Rolland in Bouleurs, the central government's mandates for municipalities to curb urban development have intensified. The French Climate and Resilience Law, enacted in 2021, sets national quantitative objectives for achieving NNLT by 2050, with an intermediate goal set for 2031. The regional level has gained new responsibilities and resources, in charge of translating NNLT objective to the sub-national level. While these objectives are widely supported, some local representatives were quick to call for a review of governance structures and planned local implementation methods. As a result, new legislative and regulatory texts were published in 2023 (Loi N° 2023-630; Décret N° 2023-1096; Décret N° 2023-1097; Décret N° 2023-1098), which grant more authority to inter-municipal bodies for developing urban strategies on a regional scale. Moreover, sparsely populated municipalities will now benefit from new urban development rights, allowing for one hectare per municipality until 2031—a policy affecting 30,760, or 88% of French municipalities. These recent developments underscore that the scales and timelines for implementing the NNLT objective are subject to ongoing review and adjustment. They also raise questions on how local stakeholders navigate this evolving context to influence land policy strategies. The NNLT goal has introduced a new urgency for government intervention and restrictive regulations in land-use discussions. This increases the pressure on municipalities to find adequate compromises between the diverging interests, not only between public and private actors, but also across different levels of government.

The Bouleurs case provides insight into the role of municipalities in implementing land policy instruments. The arising conflicts illustrate the significant power and responsibilities municipalities hold within France's decentralised land policy system, even as they struggle with limited financial, staff, and time resources. In the predominantly small municipalities across France, the intent to enforce land policy faces a strong tradition of land property rights. This requires careful adjustments to establish a public stance that balances fairness to private interests with the fulfilment of public needs. National planning mandates for municipalities can be contradictory, leading to conflicts of interest that must be resolved at the local level. This situation is particularly common on the periphery of major metropolitan areas like Paris, Toulouse, Lyon, and Pau, and reflects similar challenges in other European cities (Wandl & Magoni, 2017). In the context of the NNLT goal, peri-urban municipalities are expected to preserve agricultural and natural areas, while also accommodating high demand for housing and employment. This creates significant land pressure that must be managed by the municipalities, often leading to inconsistent and inefficient outcomes.

Historically, the demographic growth of peri-urban municipalities and the associated land consumption were not viewed as problematic by private and public actors. However, as awareness grows about the costs of development—particularly infrastructure investments required by municipalities—this perspective starts to shift. National policy goals like the NNLT objective are reinforcing this change. Yet, these policies do not provide the necessary instruments needed to find compromises at the local level. Consequently, new solutions will need to come from developers', municipalities', or landowners' initiative. France values private initiative but remains committed to public leadership. As the urgency of NNLT goals increase, particularly for small municipalities, these policy goals are becoming more pressing for local governments. The challenge of integrating and implementing overarching instruments to achieve land policy goals thus continues to be a critical task for policy makers in France.

References

Acosta, R., & Renard, V. (2018). *Urban land and property markets in France*. Routledge.

Booth, P. (1998). Decentralisation and land-use planning in France: A 15 year review. *Policy & Politics, 26*(1), 89–105. https://doi.org/10.1332/030557398782018310

Comby, J., & Renard, V. (1990). *Land policy in France, 1945–1990*. Association des études foncières.

Couleaud, N., Lenseigne, F., & Moreau, G. (2021). *La France et ses territoires*. Institut national de la statistique et des études économiques (INSEE).

Figeat, D. (2016). *Mobilisation du foncier privé en faveur du logement*. Ministère du logement et l'habitat durable.

Guelton, S. (2018). The end of the French model of land development? *Town Planning Review, 89*(6), 553–556. https://doi.org/10.3828/tpr.2018.38

Guelton, S., & de Flore, R. (2020). Les enjeux fonciers vus d'en bas: Quand les mobilisations des propriétaires privés se font citoyennes. *Métropoles, 27*. https://doi.org/10.4000/metropoles.7327

Guelton, S., Navarre, F., & Rousseau, M.-P. (2011). *L'économie de l'amenagement du territoire*. SOTECA, Belin, France. ISBN 9782916385693.

Guelton, S., & Pouillaude, A. (2023). Can the French development tax be a strategic land value capture instrument? *Town Planning Review, 94*(2), 215–239. https://doi.org/10.3828/tpr.2021.56

Hilal, M., Legras, S., & Cavailhès, J. (2018). Peri-urbanisation: between residential preferences and job opportunities. *Raumforsch Raumordn, 76*(2), 133–147. https://doi.org/10.1007/s13147-016-0474-8

Jehling, M., & Hecht, R. (2022). Do land policies make a difference? A data-driven approach to trace effects on urban form in France and Germany. *Environment and Planning B: Urban Analytics and City Science, 49*(1), 114–130. https://doi.org/10.1177/2399808321995818

Lacoere, P., Hengstermann, A., Jehling, M., & Hartmann, T. (2023). Compensating downzoning. a comparative analysis of European compensation schemes in the light of net land neutrality. *Planning Theory & Practice*, 1–17. https://doi.org/10.1080/14649357.2023.2190152

Le Bivic, C. (2018). Emprirical material based on interviews (unpublished).

Le Bivic, C., & Melot, R. (2020). Scheduling urbanization in rural municipalities: Local practices in land-use planning on the fringes of the Paris region. *Land Use Policy, 99*, 1–10. https://doi.org/10.1016/j.landusepol.2020.105040

Pichon, A. (2022). *L'ingénierie au service de la planification dans le périurbain francilien* (Master's thesis, Paris School of Urban Planning, DUI course, Gustave Eiffel University/UPEC).

Rannou-Heim, C. (2022). Populations légales au 1er janvier 2020: 12 271 794 habitants en Île-de-France (Version 75) . INSEE. Retrieved July 15, 2024, from https://www.insee.fr/fr/statistiques/6678407

Shields, K. (2023). Review of France's SAFER Land Market Interventions (Commissioned Report). Scottish Land Commission.

Wandl, A., & Magoni, M. (2017). Sustainable planning of peri-urban areas: Introduction to the special issue. *Planning Practice & Research, 32*(1), 1–3. https://doi.org/10.1080/02697459.2017.1264191

Camille Le Bivic is a researcher in spatial planning, land policy and rural development. Her approach in the fields of sociology of law and public policies, applied to local and regional spatial planning, contribute to the understanding of how actors manage and control urban development and land preservation. After a Ph.D. at Université Paris Saclay and INRAE and postdoc research at Université Gustave Eiffel in France, she is currently working on farmland preservation and changes in the French NGO FNSafer.

Sonia Guelton is Professor at University of Paris Est Creteil/Parisian School of urbanism, France and a researcher at Lab'Urba, a research team specialised in urban planning and development. She also leads a think tank on land issues Fonciers-en-débat. She works on local economics and public policies, she is specialised in land taxation and socio-economic impact of development policies. Her recent research focusses on land economics.

Mathias Jehling is senior researcher at Leibniz Institute of Ecological Urban and Regional Development (IOER) in Dresden, Germany, where he leads the research group on "Urban Structure and Policy". His focus is on geographic information in the planning context. He works and teaches (Technical University of Dresden) on urban form and institutionalist approaches to planning and land policies. He obtained his Dr.-Ing. at Karlsruhe Institute of Technology.

Caspar Kleiner is a researcher at the Leibniz Institute of Ecological Urban and Regional Development (IOER) and doctoral candidate at TU Dortmund University. With a background in architecture and urbanism he focusses on the interlinkages of land policy and urban form.

Land Policy in Germany: Waiting for the Owner to Develop

Thomas Hartmann and Fabian Wenner

Abstract Land Policy in Germany is not an independent policy domain but rather takes on different meanings across various spatial scales. At the national level, it is understood as a non-unified policy field that emerges at the intersection of building, planning, environmental, and other policies. On the local scale, it is commonly understood as the strategy of municipalities to implement land-use planning objectives by means of private law and property rights. On the national level, land policy objectives, such as the production of 400,000 housing units per year, reducing land take to 30 ha per day, and urban densification, have been established. However, these objectives are partially in conflict with each other. Moreover, on the local level, the strong property rights associated with land ownership can make it challenging to implement these policy goals. The case of *Burgwall 21* in the city of Dortmund to illustrate how land policy in Germany shapes and influences the relationship between public planning and property rights in practice. The case study explores why a plot of land with high development potential and economic value in a well-integrated urban location can remain underdeveloped for decades.

1 German Understanding of "Land Policy"

The term "land policy" translates as *Bodenpolitik* in German. In post-war Germany, the term has been used in two related yet distinct ways.[1]

At the national and sometimes state level, *Bodenpolitik* is viewed as a not formalised or unified policy field. It emerges at the intersection of various policy

---

[1] Historically, the term Bodenpolitik had been used in Nazi-Germany as well, with a geopolitical, territorial connotation along the ideology of 'blood and soil' (Bajohr and O'Sullivan 2022).

T. Hartmann (✉)
TU Dortmund University, Dortmund, Germany
e-mail: thomas.hartmann@tu-dortmund.de

F. Wenner
RheinMain University of Applied Sciences, Wiesbaden, Germany
e-mail: fabian.wenner@hs-rm.de

© The Author(s) 2025
T. Hartmann et al. (eds.), *Land Policies in Europe*,
https://doi.org/10.1007/978-3-031-83725-8_8

fields, including building and planning, environmental policy, and property taxation law. At this level, land policy is not concerned with specific plots of land but with the general policies and laws governing land use and ownership. *Bodenpolitik*, in this sense, is the overarching framework of laws and regulations governing the distribution as well as the rights and obligations associated with ownership of land. Examples of concrete political actions that fall under this definition include governmental expert commissions (such as the recent building land commission in 2019) or property tax reforms.

A second, narrower understanding of *Bodenpolitik* refers to the specific implementation of policies related to private land at the local level. This approach aims to adapt property rights according to land-use planning objectives (*Bauleitplanung*), often through private law or other means beyond traditional zoning. The extent to which municipalities implement such land policies varies widely, with different municipalities adopting different strategies. The term land policy is often used together with a qualifying adjectives, such as 'active' or 'socially just', indicating the range of different configurations that can exist at the local level. Land policy encompasses various public policy instruments, such as land readjustment, expropriation, pre-emption rights etc.

The case of "*Burgwall 21*" in Dortmund highlights how Germany's land policy, with its dual interpretations of national frameworks and local implementation strategies, shapes the relationship between public planning goals and private property rights.

2   Understanding German Land Policy: *Burgwall 21* in Dortmund

Dortmund, one of Germany's ten largest cities with nearly 600,000 inhabitants, is located in the Ruhr area, a conurbation home to approximately 5 million inhabitants as of 2023. Like the broader region, Dortmund has undergone significant post-industrial restructuring since the closure of local coal mines and steel plants around the turn of the millennium, industries which had dominated employment and city life for more than a century. The Ruhr area's urban structure is often described as a "city in between" (Sieverts, 2013), meaning that it is neither highly urbanized nor rural. Dortmund can therefore be described as a collection of suburbs around a denser urban centre with historical roots.

In the heart of Dortmund, at the intersection of *Burgwall* and *Burgtor* streets, lies a plot of land measuring approximately 3700 m² that remains almost entirely undeveloped (see Fig. 1). Known as *Burgwall 21*, this plot fulfils all land use planning conditions for construction, yet it remains 'vacant' or underutilised, Currently, the site is occupied by a parking lot, a sales kiosk, and a small grove, none of which align with its designation as a business zone in the land-use plan (*Kerngebiet*). The situation at *Burgwall 21* characterises an "economy of keeping", where land remains vacant despite its potential for development (Davy, 2006).

Burgwall 21 is a historically and economically favourable site. In 1856, Dortmund's first hotel outside the medieval city wall was built on this site. It profited from the adjacent new railway station and the demand for overnight stays it generated. Although nothing on the site today reflects its historical significance, the location has retained its economic appeal, as evidenced by the recent construction of several hotels nearby, which still profit from their closeness to the main station and the newly built Concert Hall. This highlights the area's attractiveness for construction, especially given the rising demand for housing in Dortmund over the past decade (though this demand has been decreasing since 2022) (Board of Real Estate Appraisal for the City of Dortmund, 2024).

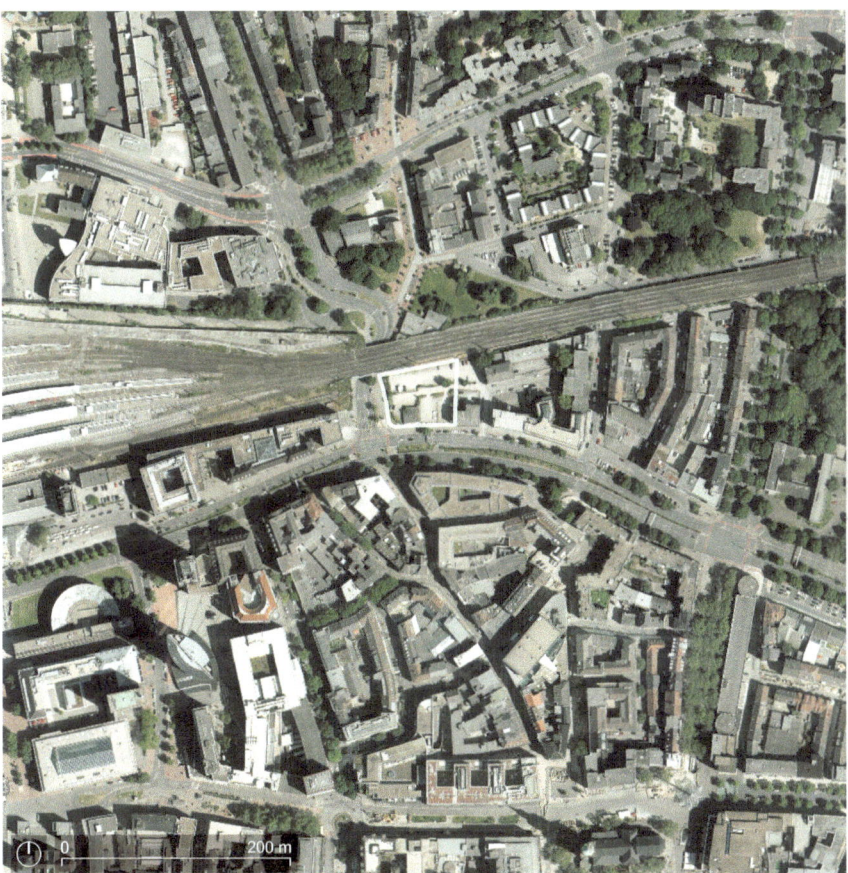

Fig. 1 Burgwall 21 in Dortmund, Germany (map: Kleiner, Jehling 2024, Aerial Imagery Provider © Esri, Land NRW, Maxar, Microsoft)

From an urban design perspective, *Burgwall 21* represents a 'gap' in an otherwise relatively coherent urban landscape characterised by a harmonious building heights, facades, and well-defined public space. Hence, *Burgwall 21* represents an almost empty plot for the cityscape. National policy also promotes inner-city development and the utilization of such vacant lots (*Baulücken*), underscoring a public interest in a different use of the plot.

In essence, *Burgwall 21* is a plot with significant developmental potential, situated in a prime area that could enhance the surrounding urban environment. The question arises: why has this potential remained untapped? To answer this, one must delve into the regulatory specifics of Germany's institutional land regime, which this case study aims to explore in-depth.

2.1 Planning Permissions

What are the relevant planning preconditions for developing *Burgwall 21*? In Germany, building permits can generally be obtained once one of the following conditions is met provided all other construction regulations are observed:

(a) Development is allowed in a binding land use plan (*Bebauungsplan*). Municipalities develop binding land use plans for specific parts of the city, ranging from single plots to entire neighbourhoods, based on the framework defined by the Federal Building Code (*Baugesetzbuch*) and its dependant ordinances. These plans typically designate the type of land use (zoning) as well as set maximum building dimensions. Once enacted, landowners within the plan area are entitled to building permissions, provided their applications conform to the plan's framework and all other relevant laws.

(b) The plot is located in an already built-up area not covered by a binding land-use plan. In many German cities, a substantial part of the urban area (around 50% in many cities) is governed by this regulation, often referred to as '§34-area' in planners' jargon, after the relevant paragraph of the Federal Building Code. It allows landowners to obtain building permission if the proposed project aligns with the characteristics of the surrounding area (in terms of use type, height, appearance, etc.). Initially introduced to accelerate post-war reconstruction, this regulation was meant to become redundant once all parts of a city had been zoned with binding land-use plans, but it has still persisted. Similar regulations exist in other European countries, such as Belgium (Hengstermann & Hartmann, 2021).

Land outside both built-up areas and binding land-use plans is, in principle, not available for new development (§ 35 of the Federal Building Code), with exceptions for a few privileged agricultural and infrastructure uses.

For *Burgwall 21*, a binding land use plan adopted in 1974 permitted the construction of a 20-story building for commercial, office, and residential use. In 2008, this plan was revised by the municipality, reducing the maximum building height to six stories. The revision was justified by new urban design principles in general and aimed to rectify the "currently unsatisfactory situation and to control future development with appropriate planning regulations" (City of Dortmund, 2008, own translation). In principle, such a downzoning could entitle the landowner to compensation, as it constitutes a seizure of a regularly vested property right. However, following constant jurisdiction, compensation cannot be claimed if construction has not commenced and the binding land-use plan in in force for more than seven years (Lacoere et al., 2023).

In conclusion, the landowners of *Burgwall 21* have the right to receive a building permit for a six-story building, provided they observe all other applicable regulations. Even without a binding land-use plan, the landowner would have the right to build on a similar scale based on §34, as the plot is located within an already built-up area.

2.2   Land Value and the Building Land Paradox

Not only is the site eligible for development, but it is also economically viable. To assess this, we can consult the reference land values for the area. Reference land values are average land zonal land values derived by statutory independent valuation boards in each municipality or district, based on recent land transactions. They have been an integral part of German planning law since its establishment, which is remarkable given that in most countries, land valuation is embedded in land taxation regulations rather than planning law (Hartmann & Maurer, 2023). In most German states, reference land values can be publicly consulted, which enhances market transparency. For example, in North-Rhine Westphalian (NRW), land reference values can be accessed through the BORIS system (*BORIS, Bodenrichtwertinformationssystem,* www.boris.nrw.de).

In 2023, the land reference value of *Burgwall 21* was 900 €/m^2, a relatively high land value for the city of Dortmund. A high land value signals demand and potential income streams, suggesting that, despite rising potential acquisition and construction costs, *Burgwall 21* is a prime candidate for development in Dortmund.

Currently, the annual property tax for *Burgwall 21* is approximately 2,500 €. This tax is calculated based on the reference land values and standardised building values. Although a recent property tax reform in Germany allowed federal states to develop their own property tax systems (with the state of Baden-Württemberg opting for a land value tax, for example), NRW has decided to retain the federal model, which is a modified version of the mixed land-and-building tax in place since 1964.

The ratio of the property tax to the overall site value is not particularly high, unlike in other OECD countries, and currently does not exert a significant pressure on landowners to develop land accordingly (Roboger, 2023). However, the government of NRW, like those of several other states, intends to introduce the option for municipalities to levy a supplementary tax for unbuilt but developable land (*Grundsteuer C*) in 2025. It is unclear whether it would apply to *Burgwall 21*, as it is already partially built-up, but could potentially shift the tax-to-value ratio and increase incentives for development. It remains to be seen whether municipal governments will adopt this new instrument, as—from a political economy perspective—they also seek the support of private landowners, an important voter base.

In conclusion, *Burgwall 21* has substantial development potential. It seems economically viable and legally permissible to build on the site. Yet, the potential remains unrealised. One contributing factor is the relatively low property taxes in Germany. This situation is not exceptional for urban land in Germany. Davy described the phenomenon of un- or underused building land as the "building land paradox" (Davy, 1996). In Germany, there is both a shortage and an oversupply of building land, as landowners may be unwilling or unable to sell or develop their land, while households who do not own land are in search of building opportunities. As a result, building land is scarce, and municipalities often designate new greenfield sites for development instead of activating existing building land potentials. This ultimately leads to inefficient land uses and less sustainable urban structures.

2.3 Housing Shortage, Densification and Reducing Land Consumption

The national land policy objectives of reducing land consumption to 30 ha per day, building 400,000 housing units per year, and at the same time promoting densification converge at individual plots of land like *Burgwall 21*.

On the one hand, the German national government has recognized a long-standing housing shortage. In 2018, the Federal Ministry, then known as the 'Interior, Building and Community', initiated the 'Expert Commission on the Long-term Provision of Building Land and Land Policy'. This commission was tasked with recommending strategies to accelerate and increase the production of building land. These recommendations led to the 2019 reform of the Federal Building Code, known as the 'Law for the Mobilization of Building Land' (*Baulandmobilisierungsgesetz*). In essence, the law strengthened the legal tools available to municipalities to compel landowners to develop their land, including lowering the conditions for pre-emption rights, accelerating planning procedures for binding land use plans for housing, and easing conditions under which a building obligation can be enacted, among other measures (Hengstermann & Hartmann, 2021).

From 2021 onwards, the new three-party coalition government—comprising of the Social Democratic Party (*SPD*), the Green Party (*Die Grünen*) and the Free Democratic Party (*FDP*)—set a target of 400,000 new housing units annually, 100,000 of which were to be in the 'affordable' sector, as stated in their coalition agreement. While the housing crisis was previously primarily framed as an issue of the rental market, leading to measures like the proposition of a rent cap (*Mietpreisbremse*), it is now increasingly framed as a land question. This shift in perspective has led to a focus on accelerating planning procedures and revising land policy instruments to address the housing shortage (Hengstermann & Hartmann, 2021).

On the other hand, land is considered a scarce environmental resource. Over the past several decades, the preservation of open space has become an important policy objective in many European countries, culminating in a no-net-land-take policy at the European level (Lacoere & Leinfelder, 2023). In Germany, one of the key milestones in this objective is the so-called '30-hectares goal', which was incorporated into its national sustainability strategy in 2002. At the time, the daily conversion of land from open space (such as natural or agricultural land) to built-up land uses or transportation infrastructure was around 120 hectares. The goal was to reduce this 'land consumption' to 30 ha per day by 2020. However, as it became clear that the target would not be met, the deadline was extended to 2030. Nevertheless, in 2023, this figure had decreased substantially to around 55 hectares (BMUV, 2023). The 30-ha goal is also enshrined in other laws, most notably in the Federal Building Code.

At first glance, the objective of conserving green space appears to conflict with the aim of building 400,000 housing units per year. Therefore, urban densification is seen by many professional planners and policymakers as a potential compromise to solve this conundrum—at least to some extent. Urban densification involves identifying unused or underused sites within the existing built-up area and focusing on their development, through strategies such as infill development, adding storeys to existing buildings, or redeveloping brownfield sites. Densification is thought to offer ecological benefits by reducing greenfield conversion as well as economic and social benefits, such as lower costs for public infrastructures and services, increased agglomeration advantages, enhanced social interaction, improved urban design.

Based on these considerations, it can be presumed there is a public interest to develop the plot on *Burgwall 21*, and the landowner's continued inaction is in conflict with this public interest. However, while strategies like the 30-ha goal are drafted at the national level, implementing land policy is largely the responsibility of the municipalities, which enjoy a high degree of independence. Additionally, it is often the private landowners who ultimately decide whether and how construction takes place. In other words, the realisation of national policy objectives 'on the ground' depends on a number of independent local actors, both private and public.

3 Actors in German Land Policy

While there are many different actors involved in German land policy, the primary private and public actors relevant to *Burgwall 21* are identified below.

3.1 Private Actors and Property Ownership

Burgwall 21 is owned by a private local developer who holds several plots of land throughout the city. However, this information is publicly available only due to media coverage and discussions on platforms such as the Deutsches Architektur-forum. In general, and unlike in many other European countries, property ownership information is not public in Germany (Hartmann et al., 2024).

Ownership, along with the rights and obligations associated with land, is regis-tered in the land cadastre (*Grundbuch*). The main purpose of the land cadastre is to guarantee a reliable basis for real estate transactions (Riedel et al., 2020) and to record land rights and obligations. Access to the land cadastre is restricted to individuals who can demonstrate a legitimate interest in the information (§12, Law on the Land Cadastre; *Grundbuchordnung*). This interest could be legal, factual, or economic. The legislator justifies this restrictive access by emphasising the need to protect landowner's economic privacy (Riedel et al., 2020). As a result, there is limited statistical information available on ownership distribution and concentration in the German land market.

Owner-occupied properties, particularly self-built single-family homes, remain the ideal housing type for most German households, as reflected in regular building land and urban development reports (*städtebauliche Berichte, Baulandberichte*) submitted by the government to the parliament. Paradoxically, despite this preference, Germany has one of the lowest homeownership rates in Europe, with only 46.5% of residential properties being self-owned (Statistisches Bundesamt, 2020). Corre-spondingly, tenant protections in Germany are relatively strong (Debrunner et al., 2024). Although the banking and mortgage system is oriented toward homebuyers, offering a range of financial products for saving to build or buying real estate (*Baus-paren*), banks are typically risk-averse regarding real estate. As a result, individual buyers often need to provide substantial advance financing—typically between 20 and 30% of the property's value—to secure a mortgage.

3.2 Public Actors and Land Policy

To understand which public authorities influence land use at *Burgwall 21*, it's necessary to examine how land policy is enacted in Germany.

As previously mentioned, the municipal level plays a crucial role in land policy for *Burgwall 21*. The right to municipal self-government is constitutionally guaranteed (article 28 of the constitutional law, *Grundgesetz*). This self-governance includes authority over land-use planning. Although city councils formally make decisions on land policy and planning, in practice, the executive branches of municipalities exercise significant independence in their decisions.

Among public actors, the most influential is the department for planning (*Stadtpla-nungsamt*), but others, such as the departments for construction supervision (*Bauord-nungsamt*), housing (*Wohnungsamt*), or municipal real estate (*Liegenschaftsamt*) also have impact on landowners' actions. The planning department, however, is the most relevant:

The planning department prepares binding land-use plans, which directly regulate what may or may not be developed on specific plots of land. These binding plans are part of a hierarchy of comprehensive plans, which include preparatory land-use plans at the municipal level and regional plans at the state level (for more information on the planning system, see Reimer et al., 2014). However, only the binding land-use plan provides enforceable building rights for landowners; the other plans are binding only for public authorities. Consequently, federal and state planning only have an indirect influence on the actual land use of *Burgwall 21*. The construction supervision department is responsible for interpreting these binding land-use plans and issuing building permits for landowners.

The housing department manages public housing stock and, in some cases, supports private residential construction. For example, the City of Dortmund's housing department proclaims that it "helps citizens [realize] their dream of the individual self-owned home" (City of Dortmund, 2023). This department plays a crucial role in incentivising landowners through subsidies and support for housing construction. Municipal real estate departments manage a municipality's real estate portfolio, including both residential and otherwise. These departments may actively participate in land markets by buying or selling land to influence or execute spatial developments, though the level of activity varies widely between municipalities.

Additionally, many municipalities have created independent urban development companies (*Stadtentwicklungsgesellschaft*). These entities, typically organised under private law but wholly owned by the municipality, operate independently of the rights and duties of public institutions. This independence allows for a high degree flexibility and autonomy in their operations.

4   Institutions of German Land Policy

To better understand the municipality's limited agency in enforcing appropriate use land use, we will explore potential options for action.

The land use of *Burgwall 21* can be influenced by the municipality through different instruments of land policy. The choice for or against a particular instrument is influenced not only by the legal framework and principles of administrative action—such as the principle of proportionality or the rule of law—but also by practical considerations such as the municipality's budget situation. In addition, political considerations may sway the choice of instruments. Generally, however, municipalities have a certain amount of discretion in most cases.

The specific use and combination of these instruments can be considered strategies of land policy (Gerber et al., 2018). We can distinguish four types of strategies (Shahab et al., 2021), with varying implications for *Burgwall 21*: active, passive, reactive and protective. These terms describe the relation between public authorities and land markets: passive land policy is supply-driven, reactive land policy is demand-oriented, active land policy is profit-orientated, and protective land policy strategy emphasizes welfare orientation. However, these four types are ideal models and are often intertwined in practice.

4.1   Passive Land Policy

In passive land policy, municipalities grant building rights for certain areas without actively participating in the market. Owners of these areas receive the planning gain and can decide independently whether they want to develop the land in accordance with the specifications (Hartmann & Spit, 2015). Passive land policy is supply-driven and represents the traditional German planning approach of "supply-oriented planning" (*Angebotsplanung*) (Krautzberger, 2010) with a reduced role of governments in land markets (Shahab et al., 2021).

In Dortmund, passive land policy was already in use during the reconstruction phase after World War II. Reconstruction after 1945 was focussed on adapting the streetscape and plot boundaries to be more car-friendly, as well as reducing plot fragmentation, which was considered progressive and economical at the time. As such, reconstruction was strongly tied to re-parcelling the land with land readjustment.

Land readjustment is a land policy instrument that reallocates plots of land in a certain area among the existing owners, ensuring their location, shape, and size are suitable for future use usually for development. Mandatory land readjustment can reallocate property rights even against the will of individual owners (Davy, 2007). In practice, however, municipalities typically adopt a cooperative approach with landowners. Land readjustment must be in the interests of the property. An important principle of land readjustment is that it should benefit landowners, requiring equal treatment regarding benefits and burdens. This necessitates transparent land valuation, illustrating the close connection between land valuation and land policy in Germany.

Land readjustment primarily serves an ordering function, helping to implement the land use plan (principle of conformity). Landowners are still free to use the land according to the plan, or leave it as it is, though they cannot construct new buildings that deviate from the plan. This helps to understand why *Burgwall 21* can remain un-, or at least under-developed.

Consistent with the passive land policy strategy, many municipalities today develop 'building land cadastres' to map potentials for inner development and densification. The Federal Building Code explicitly mentions the possibility of establishing a cadastre of underused building land, which serves as a preparatory step for municipalities to inform their planning decisions. Dortmund, for example, developed a building land cadastre in 1993 and is developing a new one in 2023. These cadastres do not impose any legal obligations on landowners but serve as tools for communication and information. Some sites included in the building land cadastre remain unchanged, such as *Burgwall 21*.

German planning law includes further instruments of land policy that the City of Dortmund could use to activate the building potential of *Burgwall 21*. One remarkable one is the building obligation. It allows a municipality to compel landowners to develop their property within a reasonable time. However, while there are many plots like *Burgwall 21* in Germany are subject to the building obligation, it is rarely ever used in practice (Kolocek, 2018). Municipalities are hesitant to enforce it due to high administrative costs and low implementation strength. If the landowner is unable to develop the land, the municipality may have to step in, which poses a financial risk. Nonetheless, the building land commission recently recommended strengthening the building obligation, and planning law has been adapted accordingly (Hengstermann & Hartmann, 2021).

4.2 Reactive Land Policy

An alternative to passive is a reactive land policy strategy. If the owner of *Burgwall 21* intends to erect additional or different buildings than those prescribed in the binding land-use plan, they can approach the municipality to negotiate new building rights. If an agreement is reached, municipality and landowner can enter into an urban development contract (§11 Federal Building Code). In exchange for increased building rights, the landowner agrees to implement certain public interest 'obligations', financed by the land value gain.

Urban development contracts were permanently introduced in planning law in 1998 (Schmidt-Eichstaedt, 2019), within the context of a 'neoliberal' political climate. Since then, municipalities have increasingly used such contracts for urban development projects, albeit with low transparency. It is estimated that large cities such as Dortmund have concluded several hundred contracts. However, German administrative law restricts the extent of developer obligations. Municipalities may only demand obligations—whether in money, land, or services that are directly related to the development project in question (*Kopplungsverbot*). This ensures that municipalities do not misuse their administrative powers against private parties.

Many municipalities have developed standard frameworks for these contracts, manifested in building land models (*Baulandmodelle*), often passed as statutes by the local councils. These standards typically require a certain percentage of affordable housing (e.g., 25–30% of the development), the construction of social infrastructure by the developer, and other conditions. The 'Munich model' for socially just land use (*Sozialgerechte Bodennutzung*) was the frontrunner in this approach. Building land models allow municipalities to take a reactive approach: when a developer approaches the municipality with a proposal, the municipality responds by promising building rights in exchange for certain developer obligations (Shahab et al., 2021). This strategy is demand-oriented and has become widespread in German planning practice since the turn of the century.

Dortmund's building land model, revised in 2022, resulted from a collaborative process involving key stakeholders in the city, including developers, housing associations, and citizens (City of Dortmund, 2021b). It stipulates that at least 25–30% of floor space in new development projects on private land must be affordable housing (City of Dortmund, 2021b). However, since a binding land use plan is already in place for *Burgwall 21*, developer obligations could only be applied if the existing plan is amended, or an exemption is granted at the developer's request.

4.3   Active Land Policy

The city of Dortmund also recognizes that affordable housing projects can be more effectively realized on municipal land rather than through urban development contracts (City of Dortmund, 2021b). Therefore, Dortmund has recently strengthened its urban development company (Dortmunder Stadtentwicklungsgesellschaft) and expanded its business model to include more active involvement in housing development (City of Dortmund, 2021a).

Active land policy, as strongly recommended by the building land commission, involves the municipality becoming an active player in the real estate market. The municipality purchases land, develops it, constructs buildings, and sells them. Recent reforms in planning law support this strategy by simplifying the application of pre-emption rights and land-use plans for housing. Some municipalities, such as the city of Ulm, have followed this strategy for decades and now own of a significant part of their territory. Ulm only enacts new binding land-use plans on land that it owns. However, there are currently no concrete plans by the city of Dortmund to purchase and develop *Burgwall 21* under municipal responsibility.

4.4   No Explicit Land Policy

So, while passive, reactive, and active land policy are possible pathways for the development of *Burgwall 21*, and the City of Dortmund is using such strategies in different projects, for *Burgwall 21* the city is not actually pursuing a dedicated strategy of land policy at all. While there have been several changes in land use planning regarding the plot over the last decades, they seem to have been entirely disconnected from the question of land ownership and disposal rights. At the same time, the city of Dortmund has employed other strategies of land policy before in other cases (e.g. a very active land policy in the case of the 'Lake Phoenix' redevelopment; Davy, 2012), showing that it is in principle able to pursue a dedicated land policy for a given plot. However, the case of *Burgwall 21*, can arguably best described with 'no explicit land policy' at all.

5   Reflection

Burgwall 21 illustrates the interplay of rights and obligations between private landowners and municipalities under the German *Bodenpolitik* as a national policy. It also highlights the opportunities available for municipalities to conduct their own *Bodenpolitik* at the local level.

In this reflection, we argue that German land policy, while encompassing aspects of the four land policy strategies (passive, reactive, active, none) discussed earlier, is shaped by four underlying ideologies. First, German land policy is property-friendly. Second, it is more crisis-proof and resilient than responsive and flexible. Third, it typically emphasises land-use regulation over visionary planning. Fourth, it is growth-oriented.

(1) Property friendliness: The property-friendly nature of German land policy can be traced back to the post-war period. After the devastation caused by extremism and dictatorship, and in opposition to the emerging competing East German communist model, lawmakers sought to stabilise the liberal democratic order. Individual property became a cornerstone of this effort, both constitutionally and politically. The belief was that property owners would favour political stability over revolution and extremism, seeing property as a means to secure individual liberty and limit state power (Andersen, 2003). This perspective was embedded in early West-German housing laws (Deutscher Bundestag, 1953) and became a fundamental part of the social market economy. The Federal Building Code (§ 1 (6) No. 2) explicitly aims to provide 'wide sections of the population' with residential property. The post-war 'economic miracle' further reinforced this property-friendly approach, offering "broad segments of society the opportunity to achieve unprecedented prosperity. After three decades of war and crises, the new order brought a sense of security and normalcy for the first time" (Thränhardt, 2021, own translation). This, combined with the elements described in this chapter—such as the non-disclosure of ownership data and compensation rules—illustrates the German legal system's enduring property-friendly orientation. *Burgwall 21* serves as a prime example of how this deep respect for, or perhaps even fear of, private property rights influences the actions of civil servants responsible for enacting land policy, especially at the local level.

(2) Resilience: The resilience of German land policy is evident in its legal reforms, or the relative lack thereof. Resilience refers to a system's ability to absorb disturbances (Holling, 1973). Major events, such as the reunification of Germany, neo-liberal pressures in the 1990s, economic crises, the climate crisis, and the recent surge in construction costs, can all be seen as disturbances to the legal system. Unlike neighbouring countries that have frequently revised their planning and building laws, German land policy has evolved incrementally. The traditional passive land policy was complemented by reactive land policy in the late 1990s and early 2000s, with more recent emphasis on active land policy. This continuity allows German land policy to be characterised as resilient, absorbing external shocks and functioning adequately under new circumstances. However, this resilience comes at the cost of flexibility and responsiveness. The inertia in the development on *Burgwall 21* over the last at least 60 years exemplifies this resilient yet inflexible land policy.

(3) Regulatory focus: While some German cities, such as Munich, Ulm, Münster, and Tübingen, have implemented local land policies that actively steer spatial development according to a vision, most municipalities prioritise regulating land use to prevent nuisances rather than realizing ambitious plans. Dortmund, for example, has demonstrated its capacity for visionary projects, such as the

creation of the artificial Phoenix lake, co-financed by land rent increases from neighbouring properties. However, many underdeveloped plots, like *Burgwall 21*, remain unaddressed by visionary land policy.

(4) Growth orientation: The instruments of land policy in the Federal Building Code are generally designed to manage of growth, a primary concern in post-war Germany, rather than urban change or shrinkage. In growth scenarios, the redistribution the planning gains allows for value capture for public purposes. However, current challenges in urban development necessitate instruments better suited to managing the existing built environment and situations of stagnating or declining land values. Recent changes to the Federal Building Code have made it easier for municipalities to enforce building obligations. However, if development on *Burgwall 21* is not financially viable for the owner, the city would need additional private-law instruments and subsidies to achieve the desired planning outcomes.

Currently, national and local land policy are not always aligned. National policy promotes housing production, aims to drastically reduce land take, and pursues high quality urban development goals. An initiative by the Federal Ministry for Housing, Urban Development, and Building to introduce a "building turbo" aims at accelerating construction by allowing zoning to be bypassed in certain circumstances. Meanwhile, reducing land take remains a key priority. However, these national ambitions must be realized at the local level, where other considerations often take precedence.

Looking ahead, German land policy faces new challenges. Rising construction costs and a shortage of labour are likely to further slow the building industry. Given these factors, radical changes in German land policy are unlikely. Instead, it is expexted to remain property-friendly, resilient, regulatory rather than visionary, and focused on steady growth. Thus, if *Burgwall 21* developed in the near future, it is unlikely to result from significant land policy intervention.

References

Andersen, U. (2003). Soziale Marktwirtschaft/Wirtschaftspolitik. In U. Andersen (Ed.), *Handwörterbuch des politischen Systems der Bundesrepublik Deutschland* (pp. 559–568). BpB, Bonn.

Bajohr, F., & O'Sullivan, R. (2022). Holocaust, Kolonialismus und NS-Imperialismus. *Vierteljahrshefte Für Zeitgeschichte, 70*(1), 191–202. https://doi.org/10.1515/vfzg-2022-0008

Board of Real Estate Appraisal for the City of Dortmund. (2024). Grundstücksmarktbericht 2024 für die Stadt Dortmund. Der Gutachterausschuss für Grundstückswerte in der Stadt Dortmund, Dortmund.

BMUV (2023) Flächenverbrauch - Worum geht es? https://www.bmuv.de/themen/nachhaltigkeit-digitalisierung/nachhaltigkeit/strategie-und-umsetzung/flaechenverbrauch-worum-geht-es. Accessed 5 June 2023

City of Dortmund. (2008). Bebauungsplan InW 104–1. Drucksache (12429-08)

City of Dortmund. (2021a). Grundsatzbeschluss zur Neuausrichtung der Dortmunder Stadtentwicklungsgesellschaft mbH (DSG): Drucksache Nr.: 21044-21. https://rathaus.dortmund.de/dosys/gremrech.nsf/0/294383CBA57A84C9C12586E70075F3D2/$FILE/VorlageVG%2321044-21.doc.pdf. Accessed May 18, 2021

City of Dortmund. (2021b). Kommunales Wohnungskonzept Dortmund 2021. https://www.dor
   tmund.de/media/p/wohnungsamt/downloads_afw/Kommunales_Wohnkonzept_Dortmund_2
   021_1.pdf. Accessed May 16, 2023.
City of Dortmund. (2023). Department for Housing. https://rathaus.dortmund.de/wps/portal/dor
   tmund/home/dortmund/rathaus/domap/times.domap.de/company.times.domap.de/!ut/p/z1/jdC
   9DoJADADgZ2FgpT3Fk7hdHIg_JDII2MWAOQ8S4AigvL5E42Cih93afG3TAkECVKf3Qq
   V9oeu0HPMT8fMs9GaM-bjHheAoguU6CsIId8ggfgL8EQKB_uk3ADKPj4FMK9yIm4H
   PmBlghBMgcCcAf08w3LEFUqXOXi8XdTb3FFArr7KVrXNrx3Le9023stHGYRgcpbUqp
   XPRlY3fWnLd9ZB8SmiqY4LFpjhUsdcJy3oA7k-DEw!!/dz/d5/L2dBISEvZ0FBIS9nQSEh/?
   p_id=64--0. Accessed May 16, 2023.
Davy, B. (1996). Baulandsicherung: Ursache oder Lösung eines raumordnungspolitischen Para-
   doxons? Zeitschrift Für Verwaltung, 21, 193–208.
Davy, B. (2006). Innovationspotentiale für Flächenentwicklung in schrumpfenden Städten. Flächen-
   management am Beispiel Magdeburgs. Magdeburg.
Davy, B. (2007). Mandatory happiness? Land readjustment and property in Germany. In Y. Hong &
   B. Needham (Eds.), Analyzing land readjustment: Economics, law, and collective action (pp. 37–
   55). Lincoln Institute of Land Policy
Davy, B. (2012). Land policy. A German perspective on planning and property. Ashgate.
Debrunner, G., Kolocek, M., & Schindelegger, A. (2024). The decommodifying capacity of tenancy
   law: comparative analysis of tenants' and landlords' rights in Austria, Germany, and Switzerland.
   International Journal of Housing Policy, 1–23. https://doi.org/10.1080/19491247.2024.2367835
Deutscher Bundestag. (1953). Entwurf eines Gesetzes zur Schaffung von Familienheimen (Zweites
   Wohnungsbaugesetz). Bundestags-Drucksacke (BT 2/5). https://dserver.bundestag.de/btd/02/
   000/0200005.pdf. Accessed May 3, 2024.
Dieterich, H., Dransfeld, E., & Voss, W. (1993). Urban land & property markets in Germany.
   European urban land & property markets (Vol. 2). UCL Press.
Fischer, R., Kleiber, W., & Werling, U. (2014). Verkehrswertermittlung von Grundstücken:
   Kommentar und Handbuch zur Ermittlung von Marktwerten (Verkehrswerten) und Belei-
   hungswerten sowie zur steuerlichen Bewertung unter Berücksichtigung der ImmoWertV, 7th
   edn. Bau und Immobilien. Bundesanzeiger, Köln
Gerber, J.-D., Hengstermann, A., & Viallon, F.-X. (2018). Land policy: How to deal with scarcity
   of land. In J.-D. Gerber, T. Hartmann, & A. Hengstermann (Eds.), Instruments of land policy:
   Dealing with scarcity of land (pp. 8–26). Routledge.
Hartmann, T., & Maurer, A. (2023). Bodenpolitik und Wertermittlung. In D. Schaper (Ed.), Aktuelle
   Herausforderungen für die Immobilienbewertung, Immobilienwirtschaft und Bodenpolitik -
   Festschrift für Wolfgang Kleiber zum 80. Geburtstag. Reguvis, Köln (pp 91–98).
Hartmann, T., & Spit, T. (2015). Dilemmas of involvement in land management—Comparing an
   active (Dutch) and a passive (German) approach. Land Use Policy, 42, 729–737. https://doi.org/
   10.1016/j.landusepol.2014.10.004
Hartmann, T., Roboger, C., Eisenhut, B., Novak, J. (2024). Land ownership transparency. A comparison of seven European countries. Hg. v. BBSR. Berlin (BBSR-online-publication). www.bbsr.
   bund.de
Hengstermann A, Hartmann T (2021) Grund zum Wohnen – Das Baulandmobilisierungsgesetz aus
   internationaler Perspektive. PND - Rethinking Planning 1. https://doi.org/10.18154/RWTH-
   2021-01677
Holling, C. S. (1973). Resilience and Stability of ecological systems. Annual Review of Ecology
   and Systematics, 4, 1–23. https://doi.org/10.1146/annurev.es.04.110173.000245
Kolocek, M. (2018). A German perspective on building obligations: Planning professionals try
   to remember. In J.-D. Gerber, T. Hartmann, & A. Hengstermann (Eds.), Instruments of Land
   Policy: Dealing with Scarcity of Land (pp. 189–192). Routledge.
Krautzberger, M. (2010). Von der Angebotsplanung zur Projektplanung: Tendenzen der jüngeren
   Städtebaugesetzgebung. RaumPlanung, 2010, 205–208.

Lacoere, P., & Leinfelder, H. (2023). Land oversupply. How rigid land-use planning and legal certainty hinder new policy for Flanders. *European Planning Studies, 31*, 1926–1948. https://doi.org/10.1080/09654313.2022.2148456

Lacoere, P., Hengstermann, A., Jehling, M., & Hartmann, T. (2023). Compensating downzoning. A comparative analysis of European compensation schemes in the light of net land neutrality. *Planning Theory and Practice*. https://doi.org/10.1080/14649357.2023.2190152

Reimer, M., Getimis, P., & Blotevogel, H. H. (Eds.). (2014). *Spatial planning systems and practices in Europe: A comparative perspective on continuity and changes*. Routledge.

Riedel, E., Volmer, M., Wilsch, H., Haegele, K., Schöner, H., & Stöber, K. (2020). Grundbuchrecht, 16th edn. Handbuch der Rechtspraxis, Band 4. C.H. Beck, München.

Roboger, C. (2023). The taxing implementation of densification: The missed opportunity of the German land value tax. *Raumforschung und Raumordnung, 81*(6), 636–647. https://doi.org/10.14512/rur.1721

Schmidt-Eichstaedt, G. (2019). Städtebaulicher Vertrag. In Raumforschung und Landesplanung Af (Ed.), Handwörterbuch der Stadt- und Raumentwicklung. Akademie für Raumforschung und Landesplanung, Hannover.

Shahab, S., Hartmann, T., & Jonkman, A. (2021). Strategies of municipal land policies: Housing development in Germany, Belgium, and Netherlands. *European Planning Studies, 29*, 1132–1150. https://doi.org/10.1080/09654313.2020.1817867

Sieverts, T. (2013). Zwischenstadt: Zwischen Ort und Welt, Raum und Zeit, Stadt und Land, 3rd edn. Bauwelt-Fundamente, 118: Stadtplanung, Urbanistik. Bauverl., Gütersloh.

Statistisches Bundesamt. (2020). *Eigentümerquote nach Bundesländern*. https://www.destatis.de/DE/Themen/Gesellschaft-Umwelt/Wohnen/Tabellen/eigentuemerquote-nach-bundeslaender.html?view=main. Accessed March 12, 2023.

Thränhardt, D. (2021). Bundesrepublik Deutschland: Entwicklung 1949–1990. *Handwörterbuch des politischen Systems der Bundesrepublik Deutschland* (8th ed., pp. 91–99). Springer Fachmedien Wiesbaden.

Thomas Hartmann is the chair of land policy and land management at the School of Spatial Planning, TU Dortmund University, Germany. His research focuses on strategies of municipal land policy, and the relation of flood risk management and property rights. He is the former president of the International Academic Association on Planning, Law, and Property Rights.

Fabian Wenner is professor for sustainable urban and transport planning at RheinMain University of Applied Sciences in Wiesbaden, Germany. His research focuses on integrated transport and settlement planning through accessibility, instruments of land policy for sustainable urban development, and digitalisation in urban planning.

Land Policy in Norway: Exploring the Boundaries of Planning Striving for Density and Car-Free Living

Terje Holsen, Andreas Hengstermann, and Helén Elisabeth Elvestad

Abstract This chapter explores the unique characteristics of Norwegian land policy, focusing on the institutional challenges posed by a planning system that seeks to balance prescriptive strategic land-use planning with increasingly detailed prescriptive and performative requirements for the content of plans. To illustrate these challenges, we present the case of *Mindemyren* in Bergen. This former low-intensity industrial and warehouse area in the southern part of Norway's second-largest city, Bergen is now being transformed into a mixed-use urban area, primarily featuring owner-occupied housing. However, the Norwegian planning system faces challenges in synchronising legally binding land-use plans across the three levels of government. This case study highlights the limited flexibility of planning authorities, as many spatial decisions are pre-determined by the constraints of sector-specific policies, including the integration of a light rail system and a fast bike route. The *Mindemyren* case underscores the difficulties planning authorities face in reconciling conflicting sector policies while pursuing sustainable urban development. A thorough understanding of Norwegian land policy is essential for addressing the complex challenges of urbanisation, transportation, and environmental sustainability in modern cities.

1 Norwegian Understanding of Land Policy

In Norway, "land policy" (translated: *Arealpolitikk*), which refers to the formal national policy on land use aimed at promoting the common interests of the nation, is closely intertwined with the principles of sustainable urban development. Key focuses include densification, multifunctional land-use, and the transformation of

T. Holsen (✉) · A. Hengstermann · H. E. Elvestad
Norwegian University of Life Sciences, Ås, Norway
e-mail: terje.holsen@nmbu.no

A. Hengstermann
e-mail: andreas.hengstermann@nmbu.no

H. E. Elvestad
e-mail: helen.elvestad@nmbu.no

© The Author(s) 2025
T. Hartmann et al. (eds.), *Land Policies in Europe*,
https://doi.org/10.1007/978-3-031-83725-8_9

former industrial areas. This understanding of land policy marked a significant shift in the early 1990s, when urban planning transitioned from an emphasis on expanding urban areas (particularly through greenfield development) to a more efficient use of urban land resources with the goal of reducing urban sprawl. The driving motive behind this shift was to curb and reverse the increasing per capita encroachment on natural resources by promoting densification. Furthermore, densification was seen as a means to reduce the demand for car-based transport. As a result, coordinated land use and transport planning—designed to be efficient and environmentally friendly— has been central to Norwegian land policy since the 1990s. At the time of this introduction, neither social nor economic sustainability was a prominent consideration in densification policy, although these aspects have since become more central to the discourse (Fig. 1).

Fig. 1 Mindemyren in Bergen, Norway (map: Kleiner, Jehling 2024, Aerial Imagery Provider © Esri, Maxar, Microsoft)

The general Norwegian understanding of land policy aligns well with the definition provided in the introductory chapter of this book. However, in Norway, the term *land policy* tends to be applied more specifically in relation to the strategic purpose outlined in the Planning and Building Act (PBA).[1] Thus, this chapter will focus on land policy instruments as they relate to public planning according to the PBA. Moreover, *Arealpolitikk* should be understood both as the process of designing and adapting plans, and as the process of implementing them, although greater emphasis is typically placed on the political factors governing plan design.

According to Statistics Norway, nearly 82% of the population (more than 76% of households) now live in owner-occupied housing (Statistic Norway, 2023). Notably, under the Planning and Building Act, housing affordability is not considered part of Norwegian land policy. Furthermore, affordable housing and publicly allocated rental housing make up a smaller share of the housing sector in Norway compared to most other European countries (Aarland & Sørvoll, 2021). Over the past century, Norwegian housing policy has primarily focused on facilitating the construction of owner-occupied housing, largely by offering highly subsidised plots of land to non-profit housing associations. This practice ended with the adoption of the densification policy, partly because many Norwegian municipalities own little land within city limits, aside from areas designed for public infrastructure. As a result, housing affordability has played a relatively minor role in Norwegian land policy in recent decades, though it has become a more prominent issue on the political agenda in recent years (although perhaps not as implemented policy). Due to long-standing policies promoting homeownership and the more recent densification approach, it has now become very costly for public authorities to reintroduce policies for constructing affordable housing.

Norwegian planning legislation exhibits clear path dependency (Holsen, 2021), defined as the influence of past decisions limiting the range of present choices (Raad-schelders, 1998). This can be seen in the progression of land policies from the first nationwide planning legislation in 1924 to the current PBA, adopted in 2008. Key amendments to the planning laws were enacted in 1965 and 1985, creating a reactive sequence (Mahoney, 2000) that explains the trajectory leading to the current land policy framework under Norwegian planning legislation.

2   Resource

Norway is sparsely populated with a widely dispersed settlement pattern across a large land area. It has a population density of only 17 inhabitants per km², with built-up areas covering just 1.8% of the land. The majority of the population (82.7%) lives in urban areas surrounded by agricultural land, yet only 3.5% of Norway's land is used for agriculture. Furthermore, 51.5% of their land is neither arable nor

---

[1] Act of 27 June 2008 No. 71 relating to Planning and the Processing of Building Applications (the Planning and Building Act).

suitable for grazing (Statistic Norway, 2023). This presents two major challenges for Norwegian land policy: long transport distances and limited agricultural land.

Historically, Norway has relied heavily on natural resources, transitioning from agriculture, forestry, fisheries, mineral extraction, and industrial hydropower to industries such as aquaculture, renewable energy, and oil and gas exports. Consequently, Norwegian land policy has traditionally focused on managing the rights to and the use of natural resources. As in many countries, land policies are managed through a combination of private laws that regulate land ownership and public administrative laws that govern land use, regardless of ownership.

An example of the challenges facing Norwegian urban densification policy can be seen in the transformation of the *Mindemyren* district in the city of Bergen, the country's second-largest city located on the west coast (see Fig. 1). Two primary challenges can be identified: determining which areas are suitable for densification, and how to achieve it—focusing on reducing land consumption and minimizing car-based transportation while establishing socially sustainable communities.

Mindemyren is a key example of urban transformation, a central approach to the densification currently taking place in several Norwegian cities. Unlike strategies such as consolidation (infill on individual plots within the existing urban fabric) or expansion (building on green or open areas within the city), which have not been prioritised by Norwegian planning authorities (see Fig. 2), densification by transformation is the dominant approach. This involves redeveloping former harbour, industrial, warehouse, or low-density residential areas into mixed-use. densified urban areas with a significant share of owner-occupied housing.

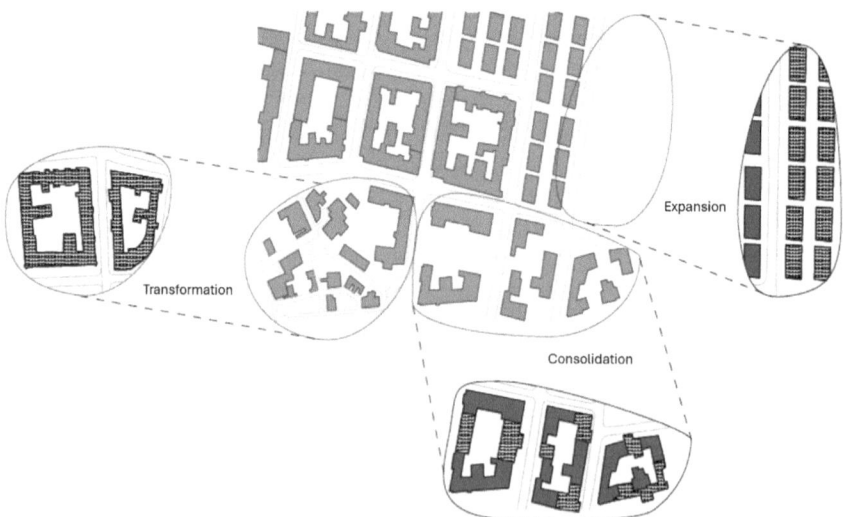

Fig. 2 Morphologically archetypical development patterns in compact city development: consolidation (Konsolidering), transformation (Transformasjon), and expansion (Expansjon) (Røsnes, 2021)

The *Mindemyren* area covers roughly 500 acres and consists of around 40 larger and smaller properties. It has an estimated development potential of approximately 800,000 square meters, with a projected completion timeline of 30 to 40 years. This transformation illustrates key features of Norway's planning system, which often involves balancing conflicting interests and managing tensions in the development process.

The need for *Mindemyren*'s transformation was first highlighted in the revision of the land-use element of Bergen's municipal master plan (*Kommuneplanens arealdel*, KPA) from 2006 to 2017 (KPA 2006). The plan emphasised the need for a gradual transformation and densification of the industrial and warehouse-dominated area with low floor area ratio[2] (FAR) and site-coverage (CVG)[3] into a mixed-use district with office, commercial, and public-oriented businesses. The plan also projected the potential to create 27,000 jobs in the area, with housing comprising 10% FAR share of housing.

A key aspect of the plan was the establishment of a high-capacity public transport network, specifically light rail. However, while the need for such transportation infrastructure was recognized, it was not initially included in the transformation plan. Large public investments would be required, and while extending the light rail network into Bergen is high on the political agenda, other sections were prioritized ahead of *Mindemyren*.

The transformation process for *Mindemyren* was formally initiated in 2008 with an area zoning plan, as existing zoning regulations did not permit the necessary changes in land use or density increases. The area zoning plan, adapted in 2014, built on the goals of the 2006 municipal master plan, emphasising the development of workplace-intensive and public-oriented businesses. It also included an updated housing target of 13% FAR for residential use, with a political goal of increasing this to 20%.

The assumed housing potential in *Mindemyren* is critical to understanding Norwegian land policy. Commercial housing development for owner-occupancy, the lion's share of all residential development in Norway, is significantly more profitable than other types of commercial real estate. This profitability could allow the municipality to capture more of the development value. While Norwegian law only allows for cost recovery of necessary technical infrastructure investments, negotiations often hinge on the developer's potential profit, meaning a higher share of housing could lead to increased value capture by the municipality.

When the area zoning plan was approved by the city council in 2014, it aimed to transform *Mindemyren* into an attractive, dense, urban business area with residential units, featuring an urban grid block structure that could accommodate a total of 22,000 jobs and 1400 residential units. The plan established a clear spatial structure, defining public spaces such as streets, squares, public transport corridors, and parks.

---

[2] FAR is an expression for the total floor area of buildings divided by the area of the parcel on which they are located.

[3] CVG is an expression for the area covered by buildings divided by the area of the parcel on which they are located.

It divided the area into 25 sub-areas, all zoned for mixed-use development, including retail, services, housing, offices, hotels, community facilities, and green spaces, while excluding manufacturing and warehouses.

The zoning regulations also set bulk parameters, such as FAR, building heights, and setback lines, for each sub-area. Retails could account for up to 20% of FAR, with at least 60% reserved for other commercial uses, and a minimum of 20% FAR designated for housing. However, some sub-areas allowed up to 60% FAR for housing, while others limited it to 10%.

Car parking was planned to be consolidated in underground facilities,[4] shared by multiple sub-areas, although specific locations for these facilities were not designated. Additionally, the area zoning plan did not allocate specific places for community facilities stating that the needs would be addressed through subsequent detailed zoning plans. It was estimated that the increase is school capacity required by the 1400 new housing units could be accommodated by existing schools in the area, so no land was reserved for new schools in the zoning plan.

When adopting the legally binding area zoning plan, local politicians questioned whether the plan would offer sufficient incentives for commercial real estate development. In response, the city council stated that future detailed zoning plans could consider higher FAR and increased building heights, indicating that they would welcome such proposals.

The city council's statement regarding higher FAR and building heights stemmed from an earlier decision to prioritise a different light rail section over the one at *Mindemyren*. Concerns arose that this shift might lead to reduced investments at *Mindemyren*. To address this, it was assumed that the potential for increased profits might still entice developers to invest, despite the lack of capacity in the public transport network.

Subsequently, due to persistent political disagreement about the implementation of the prioritised light rail section it was nevertheless decided to proceed with the *Mindemyren* section. This decision did not alter the city council's statement about possible increases in FAR. Consequently, several larger developers positioned themselves by acquiring real property in *Mindemyren*, mainly with the intention constructing housing. The zoning plan for the light rail was approved in 2017, and the *Mindemyren* section was opened in 2023.

The 2014 area zoning plan marked a milestone in the *Mindemyren*'s transformation. In the revised municipal mater plan's land-use element (KPA 2018), *Mindemyren* was designated as an "urban densification zone", though the residential share of the FAR was no longer defined (Fig. 3). This change can be attributed to property acquisitions made by housing developers following the adoption of the area zoning plan. However, more important are the expectations for increased FAR and building heights, and the decision to proceed with the *Mindemyren* light rail link due to political disagreements about the originally prioritised section.

---

[4] Underground parking supports sustainable urban developmen and urban planning regulations in many Norwegian cities encourage or require the inclusion of underground parking to manage urban density and maintain aesthetic and environmental standards.

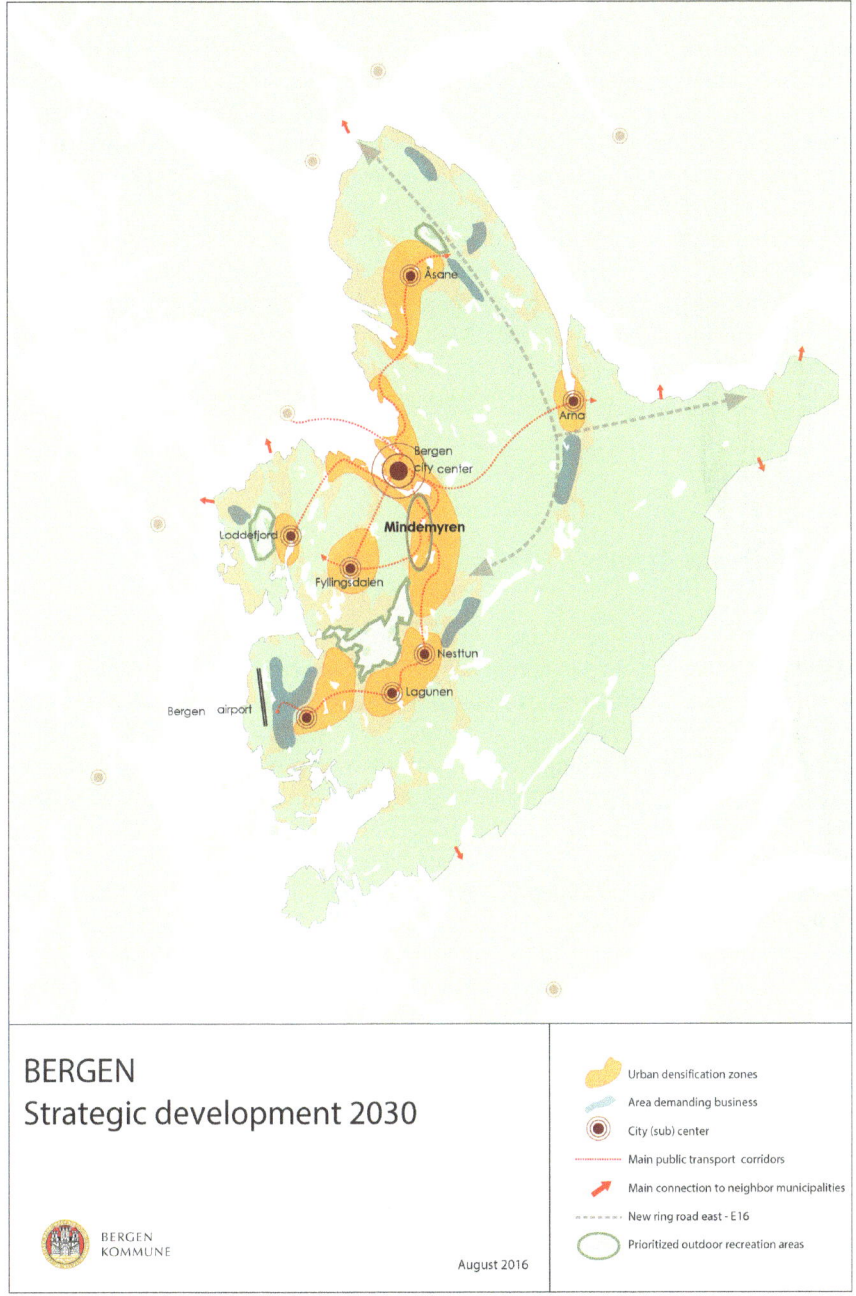

Fig. 3 Edited facsimile from Bergen municipal master plan 2018 ("KPA 2018") identifying strategic land use development. Mindemyren area is located south of Bergen city center, with the new light rail line and designated urban densification zone

As of 2024, plans are in place for the construction of up to 3,500 new residential units in *Mindemyren*, with housing constituting approximately 50% of the FAR. The first detailed zoning plan for *Mindemyren* (covering sub-areas S8 and S10) was approved in the summer of 2022, with detailed zoning plans for other major sub-areas in the pipeline for city council approval. It took more than 15 years from the initiation of the transformation of *Mindemyren* to the approval of the first detailed zoning plan. While the extended timeline is not an issue in itself, it becomes problematic when all planning levels are legally binding, as is the case in Norway. Both the land-use element of the municipal master plan and the area zoning plan were based on assumptions that had changed by the time the detailed regulation plan was drawn up. The focus had shifted from mixed-use, mainly commercial with some housing, to an urban development project emphasising significant residential development.

3   Actors

3.1   Relevant Actors in the Case

Norwegian land policy, particularly its focus on densification, heavily depends on private real estate developers and privately initiated detailed zoning plans. This is especially evident in urban transformation projects, where existing urban areas are restructured in terms of land use and density. Consequently, the principle of conformity within Norway's planning hierarchy—from overarching plans down to detailed zoning—is challenged by the fact that anyone has the right to propose detailed zoning plans. This principle was introduced into Norwegian planning legislation as early as 1924, acknowledging the need for commercial developers in urban planning.

The *Mindemyren* case highlights several issues. Despite public authorities' visions for transformation expressed in the KPA 2006 and the 2014 area zoning plan, actual construction initiatives originated with private developers. No development occurs unless developers consider a project financially viable. This creates a dynamic where developers, primarily motivated by financial returns, must work with planning authorities, who prioritise sustainable development and the public interest. The result is a negotiation-driven process, where different objectives are weighed against each other.

However, the range of actors is notably more complex, encompassing diverse public and private entities with different goals. Coordinating these actors to follow plans and their underlying objectives has proven difficult. This complexity is further exacerbated by how the municipality of Bergen structured the planning for *Mindemyren*. Key decisions regarding social and technical infrastructure, which could have been addressed in the area zoning plan, were postponed to subsequent detailed zoning plans. This lack of early infrastructure planning creates economic challenges for developers, who may try to shift these costs onto neighbouring projects, creating incentives to avoid bearing the financial burden themselves.

Furthermore, public authorities have pursued relatively distinct and sometimes conflicting policy goals at *Mindemyren* due to fragmented political authority. This fragmentation exists both between regional public sector entities and within the Municipality of Bergen. By law, the municipal council holds political responsibility for land-use planning within its jurisdiction. However, regional entities maintain sector responsibilities, often influencing municipal plans. For example, the Norwegian Public Roads Administration has played a significant role in planning and constructing the fast cycling route through *Mindemyren*. The municipality has also divided responsibilities among its various units: one unit manages the light rail, another handles blue-green infrastructure, and a third oversees kindergarten capacity, among other responsibilities.

3.2   Relevant Actors in Land Policy-Based Interventions

The constellation of actors in the *Mindemyren* case represents the four essential actors in Norwegian land policy: (1) politicians, (2) planners, (3) property developers, and (4) the population. Each actor has a distinct role, primarily shaped and guided by planning legislation.

Norway is a unitary state with three hierarchical administrative levels, each with directly elected councils. The Norwegian Parliament (*Stortinget*) is responsible for defining national land policy, while the government (*Regjeringen*) is tasked with its implementation. The government governs strategically through four-year national expectations, national policy guidelines, standards, and instructions to sector authorities regarding their specific responsibilities.

Subsidiarity and local governance are guiding principles in Norwegian public administration. Municipal councils, as the third level of governance, are largely responsible for the day-to-day political decisions on land-use planning. They also hold the authority to approve statutory land-use plans. The Planning and Building Act (PBA) mandates the preparation of national expectations every four years (PBA section 6–1). Local and regional authorities are expected to align their planning with these national expectations and guidelines. Sector authorities may object to municipally adopted legally binding land-use plans if they concern significant sectoral issues, thereby escalating descision-making to a higher administrative level.

Traditionally, land-use planning has been a public task in Norway, managed by municipal planners and supervised by sector authorities. In this context, planning legislation has created a system where public planning authorities are responsible for land policy coordination, particularly in the pursuit of sustainable development and densification. However, commercial enterprises, particularly in property development, play an increasingly central role in implementing these policies.

This shift highlights the differing perspectives of the actors involved. Politicians and public sector planners are responsible for managing the community's need for sustainable development, developers are to pursue their capital interests. The authorities aim for sustainable urban development while developers focus on maximising financial returns. Public authorities view property development as an important tool for achieving urban sustainability, whereas developers see sustainable urbanity development as necessary but secondary to their capital gains.

The evolving role of planners reflects these changes. In recent decades, a growing private sector of planning consultants has emerged, challenging the traditional role of planners. This shift has implications for planning education, which increasingly needs to incorporate property development into the curriculum. These developments also affect public participation in the planning process. Some property developers have viewed public participation as an obstacle, yet the PBA emphasises the importance of ensuring public involvement, regardless of who is leads the planning process and design. Recently, there has been a growing recognition among developers that public participation can add value, particularly in shaping projects that are more attractive and have greater sales potential. This trend reflects the increasing emphasis on social sustainability in planning practices.

4 Institutions

The *Mindemyren* case provides a practical example of how Norwegian land policy institutions operate in complex urban development settings. The formal planning system involves three administrative levels: the local (municipal) level, the regional level, and the national (state) level. Furthermore, sectoral considerations, particularly regarding infrastructure and transportation, are also crucial (Fig. 4).

The local level is the most relevant for actual development, with the municipal master plan (*kommuneplan*) at its core. This plan consists of a strategic section

Fig. 4 Institutional elements of Norwegian land policy. *Source* The authors

addressing social aspects (*kommuneplanens samfunnsdel*) and a legally binding land-use section (*kommuneplanens arealdel*). In *Mindemyren*, the master plan called for transforming a low-intensity industrial and warehouse area into a densely developed, mixed-use urban grid block structure. Beneath the master plan are two additional layers of legally binding zoning: the area zoning plan (*områdereguleringsplan*) and the detailed zoning plan (*detaljreguleringsplan*). The former plan provides coordinated and relatively detailed guidelines for a designated district or development area but is optional. In contrast, detailed zoning plans are mandatory, forming the core of land-use regulation by specifying guidelines for individual construction projects. These elements within the municipal planning system ensure a structured and legally binding approach to urban development and land use within Norway's municipalities.

For *Mindemyren* the legally binding land-use section of the master plan was first followed up by an area zoning plan, which defined the overall transport infrastructure and 25 sub-areas for development. These were then refined through detailed zoning plans which govern land use for each sub-area. Broadly speaking, *Mindemyren*'s planning adhered to the principle of hierarchical conformity, following the guidelines set out in KPA 2006, the 2014 Area Zoning Plan, and the detailed zoning plans for the 25 sub-areas.

The municipality of Bergen's regulations are influenced by broader regional and national priorities, particularly the regional planning strategy (*regional planstrategi*), which coordinates spatial challenges that cross municipal boundaries. The national Planning and Building Act (*Plan- og bygningsloven*) provides a general framework for these regulations. While regional and national influences are often indirect in shaping specific projects like *Mindemyren*, sectoral requirements—particularly in transport planning—must be accounted for. In this case, transport authorities prioritised the main bicycle route and the light rail, which were designed primarily to facilitate faster transportation through the area, rather than improve internal cycling or public transport access within *Mindemyren*.

In summary, Norwegian land policy operates with a comprehensive regulatory planning system under a unitary state, with relatively strong municipalities, weak regional bodies, and powerful sectorial interest (Fredricsson et al., 2013; Kule & Røsnes, 2010).

4.1   The Role of Private Actors and Their Influence on Land Policy

Although Norwegian land policy is state-based, private actors, particularly commercial developers, play a significant role in shaping the system through their initiatives. Under the PBA § 12–11, anyone, including developers, has the right to submit proposals for detailed zoning plans, which municipalities are required to consider. Building permits must align with detailed zoning plans.

In theory, the Norwegian planning system functions as a prescriptive framework based on hierarchical conformity (Faludi, 1989), where each plan follows a deontic "should" principle (Kalbro et al., 2010). However, the conformity principle is often overridden by a "shall" principle, meaning that the most recently adopted plan takes precedence, unless explicitly stated otherwise. The possibility of promoting private detailed zoning plans further shifts planning practice towards a more discretionary, development-led, and performance-oriented approach than the formal institutional framework suggests.

4.2 Plan Implementation, Land Value Capture, and Land Tenure Inertia

While Norwegian land policy has traditionally focused on designing and adopting plans, policies relating to plan implementation has long been a key policy concern. Two main issues have been addressed: instruments for cost recovery through development agreements and mechanisms to speed up implementation in areas with challenging land tenure.

Regarding land value capture, Norwegian land policy draws a clear distinction between capturing betterment (unearned increment) and cost recovery. The former is not permitted. The latter is allowed through development agreements, but only for technical infrastructure deemed necessary by a legally binding land-use plan (*kommuneplanens arealdel, områdereguleringsplan, detaljreguleringsplan*). Social infrastructure such as schools, kindergartens, etc., can be a prerequisite for other urban development according to a legally binding spatial plan, although cost recovery for this through a development agreement is not permitted.

Land tenure inertia, particularly in urban transformation and densification projects, has long been recognised as a challenge. Various approaches have been explored to accelerate implementation. The PBA provides two instruments addressing land tenure inertia: (1) requirements for cooperation between landowners when private actors initiate detailed zoning plans, and (2) the municipality's right to demand a redistribution of the net value increase from property development (after deducting costs) among landowners in a designed transformation area. If landowners cannot reach a voluntary agreement, the Land Consolidation Court can oversee land readjustment, redistributing land values and development rights to facilitate transformation.

This system allows landowners whose properties are designed for public purposes (e.g., parks or access roads) to receive development rights for other properties in the zoning plan area. However, a unique feature of Norwegian land readjustment is that it is governed by the Land Consolidation Court rather than being integrated into public planning. The court can only be asked to conduct land readjustment after a legally binding zoning plan is adopted.

Despite its promise, land readjustment—particularly the redistribution of net value increase—has yet to play a role in the *Mindemyren* case. Although rural land consolidation has a rich history in Norway, adapting this tool for urban settings has been slow. Norway is unique in placing land readjustment entirely within the court system (Elvestad & Sky, 2021), which has limited its practical application (Hoddevik & Holsen, 2024). Thus, municipalities have been reluctant to use this instrument, leaving Norwegian land policy with few effective tools to resolve land tenure inertia. The absence of land readjustment in the *Mindemyren* case exemplifies this situation.

4.3   Path Dependency of Norwegian Land Policy

The evolution of Norwegian land policy has followed a path shaped by significant historical shifts. From its early beginnings in the nineteenth century, when land-use legislation focused on controlling the built urban form through bulk planning (e.g., volume control, building heights, and setback lines), the focus has gradually shifted towards contemporary concerns, particularly sustainable development. Early zoning restrictions were enforced through negative covenants used by property owners to safeguard land value (Elvestad & Holsen, 2020). These covenants were crucial in shaping suburban development patterns, particularly in Oslo.

The first nationwide Norwegian planning legislation introducing land-use zoning was adopted in 1924. It was driven by the need to manage urbanisation linked to industrial growth, which was based on the exploitation of natural resources. This was expected to lead to the creation of numerous smaller towns across the country. Learning from the urbanisation challenges of the nineteenth century, this legislation emphasised comprehensive and well-managed urban development.[5]

In 1965, the adoption of a Building Act marked a shift towards social democratic welfare policies, seeking to reduce regional disparities, promote full employment, and ensure economic growth and equitable income distribution (Brofoss, 1966). This act introduces the hierarchical structure of regional, general, and zoning plans, a structure that largely remains in place today.

By 1985, the Act was influenced by international liberal trends and growing environmental awareness. The focus shifted from greenfield urban expansion to compact city development. This approach was reinforced in 1992, when a white paper[6] called for a transition towards densification and compact urban forms (Hanssen & Hofstad, 2013; Hofstad, 2012; Mouratidis, 2019; Skjeggedal et al., 2003). The 1985 Act retained the earlier planning principals but expanded requirements for municipal and county municipal planning.

---

[5] Indstilling til Lov om bygningsvæsenet [Proposal to the Building Services Act], 1919.

[6] Stortingsmelding [Report to the Storting (white paper)] nr. 31 (1992–1993) Den regionale planleggingen og arealpolitikken [The Regional Planning and Spatial Policy].

Norwegian densification policy is grounded in environmental sustainability and the perceived reality (North, 2005) that urban sprawl is unsustainable. Densification, therefore, became a solution to reduce land use and transportation demands. The national densification policy has evolved as an institutional response to these concerns, positioning densification and compact urban development as key strategies for reducing land take and curbing car-based transport (Schmidt, 2014). Social and economic sustainability, however, were not major considerations when densification was formulated as land policy.

The transition to densification as the central principle of land policy was justified as a way to reverse the trend of increasing per capita use of natural resources. Higher density, co-location, and multi-use development, along with transport system reforms, were intended to reduce encroachment on nature and reliance on cars. The land policy was further reinforced by a 1996 white paper, which emphasised biodiversity and farmland protection as equally important components of the densification agenda. This aligns with the broader goals of compact urban development seen globally, which focus on high density and multi-use (Burton, 2002).

The current PBA of 2008 formalised these principles further. It mandates municipalities and county authorities to engage in comprehensive community planning with an emphasis on sustainable development. The Act's purpose clause[7] explicitly prioritises "sustainable development in the best interests of individuals, society, and future generations". This goal is supported by the National Expectations for Regional and Municipal Planning, adopted every four years by the government.[8] The most recent 2023 National Expectations have reinforced the focus on climate adaptation, farmland protection, and governance. Climate-related aspects, such as protection against natural hazards and ensuring food self-sufficiency through farmland preservation, have gained prominence. Although social sustainability and governance are mentioned, they receive relatively less emphasis, focusing on diverse housing markets, socio-demographic diversity, inclusion, and transparent planning processes.

In summary, Norway's contemporary planning framework is underpinned by a commitment to environmental sustainability, efficient land use, and resilience against natural threats, in line with broader global trends towards compact urban development. However, while social and governance aspects are acknowledged, they remain secondary in the overall agenda. Norwegian land policy has evolved in a path-dependent manner, shaped by the country's social-liberal growth and welfare policies. This approach has resulted in sustainability challenges that are now addressed through the statutory goal of sustainable development, understood primarily as environmental sustainability. This goal continually is refined and operationalised through the National Expectations updated every four years.

---

[7] Plan- og bygningsloven [The Planning and building Act] 2008 (hereafter PBL) § 1–1.
[8] Cf. PBL § 6–1.

5 Reflection

In conclusion, the case of urban densification in *Mindemyren* highlights the complexity of Norway's land policy within its intricate institutional framework. Despite Norway's vast territory and modest population, land is a scarce resource, making land policy closely tied to sustainable urban development, particularly the creation of dense, car-free neighbourhoods. The challenges of urban densification can be broadly categorised into two key areas: identifying suitable locations for densification and effectively implementing projects that reduce car dependency while promoting socially sustainable communities. The *Mindemyren* case exemplifies the transformation of a low-intensity industrial and warehouse area into a mixed-use, densified urban zone with owner-occupant housing. However, this transformation is far from a straightforward process, involving a wide range of public and private actors whose differing interests result in a complex and dynamic planning process.

The case study underscores the significance of legal frameworks, the overlapping responsibilities among authorities, value conflicts between different objectives, and the dislocation of objectives throughout the planning process. Legal complexity arises from the presence of multiple legally binding land-use plans across three governmental levels, which complicates the integration of planning and sector policy requirements. Overlapping responsibilities among different authorities further adds to these challenges, particularly in coordinating sector-specific spatial planning with local spatial initiatives. Value conflicts emerge when parallel public planning processes intersect with transformation projects, often leading to misalignment and dislocation of objectives.

While densification characterises contemporary urban and suburban development, the existing legal framework and planning system do not always align with the practical realities of planning and construction practices. Private interests, particularly those linked to property development, frequently initiate planning processes, with economic incentives often taking precedence over public interests. The challenge lies in coordinating the varying economic and institutional priorities of stakeholders.

The case study demonstrates that the transformation of *Mindemyren* is shaped by a complex institutional framework, comprising exogenous conditions, objectives, decision-making processes, and the actions of diverse actors. Understanding the outcome of these interactions requires a clear grasp of how exogenous conditions affect the positions and actions of stakeholders, as well as the broader implications of these actions.

Four main findings can explain the lengthy 15-year process to reach construction-ready land-use plans at *Mindemyren*: (1) legal complexity, (2) overlapping responsibilities, (3) value conflicts, and (4) dislocation.

5.1 Legal Complexity

Two primary sources of legal complexity are evident. First, the planning system's structure, which includes legally binding land-use plans at three geographical levels, introduces significant complexity. Second, this complexity is compounded by how sustainability is interpreted under the Planning and Building Act's purpose clause. The lack of a clear hierarchy of goals means that "everything" is treated as equally important and must be addressed simultaneously. Navigating a prescriptive framework of guidelines to secure approval for detailed zoning plans is therefore highly challenging.

At *Mindemyren*, both aspects of legal complexity are visible, particularly at the detailed planning level. The process of adopting detailed zoning plans for the 25 sub-areas within the area zoning plan highlights the challenges of balancing diverse requirements. In the absence of a clear goal hierarchy, quantitative requirements often take precedence over qualitative judgments, leading to a lowest common denominator approach.

5.2 Overlapping Responsibilities

The Norwegian Planning and Building Act assigns municipalities responsibility for local coordination of spatial planning. Simultaneously, sector-specific spatial planning is carried out under separate sectoral legislation. This creates a two-fold coordination challenge: one involving the division of responsibilities under the Planning and Building Act, and another concerning the relationship between sectoral and local spatial planning.

At *Mindemyren*, overlapping responsibilities have had significant impact. For instance, the Norwegian Public Roads Administration was responsible for planning the main bicycle route, while other municipal units were tasked with constructing the light rail and overseeing the structural transformation of the area. One unit managed land use and development patterns through statutory planning, while other units handled infrastructure projects including waste management, pedestrian and bicycle paths, local roads, public spaces, squares, and parks. These overlapping horizontal and vertical responsibilities have complicated the planning process at *Mindemyren*.

5.3 Value Conflicts

The Planning and Building Act mandates municipalities to ensure coordination and cooperation among sector authorities, as well as between state, regional, and municipal bodies, private organizations, and the general public. Regional and sectoral authorities have a "duty to participate" in planning when it "affects their field compe-tence" by "providing the municipalities with information that is relevant to their plan-ning" (PBA section 3-2). However, the law lacks a clear mechanism for enforcing coordination.

At *Mindemyren*, value conflicts have arisen, particularly in sector-oriented infras-tructure planning. For example, the fast cycling route and the light rail projects were designed to increase public transport options, with a focus on improving connections between suburbs and Bergen's city centre. However, these projects often forced trans-formation efforts in the 25 sub-areas to conform to rigid infrastructure guidelines, rather than allowing infrastructure to adapt to the transformation goals.

5.4 Dislocation

Dislocation refers to the misalignment or displacement of objectives from their expected course. At *Mindemyren*, dislocation has occurred both procedurally and substantively. Procedurally, there has been a misalignment between the overarching goals set in earlier plans (such as KPA 2006 and the 2015 area zoning plan) and the detailed zoning plans for individual sub-areas. Substantive dislocation can be seen in the clash between sectoral interests and planning levels, often leading to conflicts between objective of public infrastructure projects and transformation goals. The earlier-mentioned value conflicts are examples of substantial dislocation. Procedural dislocation can be exemplified by the changes that occurred in the plans' objectives during the planning of the *Mindemyren* transformation. There has been an ongoing development where goals set for current detailed zoning plans only to a small extent align with overarching goals for KPA 2006 and the 2014 area zoning plan.

For instance, the area zoning plan was intended to create an "urban neighbour-hood, not a business park" with "good conditions for city life, varied public spaces, and aesthetic quality." Yet, social infrastructure such as schools, kindergartens, and sports facilities were not included, likely because the plan was viewed primarily as a business development framework. Developers, seeking to maximise the economic potential of their plots, further contributing to dislocation.

5.5 Conclusion

In summary, the urban densification case of *Mindemyren* highlights the complexity of Norway's land policy, emphasizing the challenges of sustainable urban development in a country with extensive land but limited availability for urban use. Key challenges include identifying appropriate densification areas and effectively implementing projects to create socially sustainable, car-free neighbourhoods. Legal complexi-ties, overlapping responsibilities, value conflicts, and dislocation of objectives, all contribute to the prolonged timelines for construction-ready land-use plans.

References

Aarland K, Sørvoll J (2021) Norsk boligpolitikk i internasjonalt perspektiv. En sammenligning av boligmarkeder og boligpolitikk i sju europeiske land, 211. NOVA, OsloMet, Oslo.

Brofoss E (1966) *Bosetting og lokaliseringspolitikk*. Byen og samfunnet, Pax.

Burton, E. (2002). Measuring Urban compactness in UK towns and cities. *Environment and Planning. b, Planning & Design, 29*, 219–250. https://doi.org/10.1068/b2713

Elvestad, H. E., Holsen, T. (2020). Negative covenants and real-estate developers' modus operandi: The case of suburban densification in oslo, Norway. *Town Planning Review, 91*. https://doi.org/10.3828/tpr.2020.18

Elvestad, H. E., & Sky, P. K. (2021). From rural to urban land consolidation—An analysis of recent changes in Norwegian land consolidation. Working Paper No. 01/21, Norwegian University of Life Sciences (NMBU), Centre for Land Tenure Studies (CLTS), Ås.

Faludi, A. (1989). Conformance vs. performance: Implications for evaluation. *Impact Assessment, 7*, 135–151. https://doi.org/10.1080/07349165.1989.9726017

Fredricsson, C., Smas, L., Damsgaard, O., Harbo, L., & Wimark, T. (2013). En granskning av Norges planeringssystem: Skandinavisk detaljplanering i ett internationellt perspektiv. Stockholm.

Hanssen, G. S., & Hofstad, H. (2013). Compact city policies in England, Denmark, the Netherlands and Norway. *Norsk Institutt for by- Og Regionforskning, NIBR-Rapport, 2013*, 30.

Hofstad, H. (2012). Compact city development: High ideals and emerging practices. *European Journal of Spatial Development, 10*, 1–23. https://doi.org/10.5281/ZENODO.5139751

Holsen, T. (2021). A path dependent systems perspective on participation in municipal land-use planning. *European Planning Studies 29*. https://doi.org/10.1080/09654313.2020.1833841

Kalbro, T., Lindgren, E., & Røsnes, A. E. (2010). «Naer utakt» Plan-og bygningslovreformer i Norge og Sverige. *Kart Og Plan, 70*, 27–45.

Kule, L., & Røsnes, A. E. (2010). Planning control on the Northern European periphery. *European Planning Studies, 18*, 2027–2048. https://doi.org/10.1080/09654313.2010.515821

Mahoney, J. (2000). Path dependence in historical sociology. *Theory and Society, 29*, 507–548. https://doi.org/10.1023/A:1007113830879/METRICS

Mouratidis, K. (2019). Compact city, urban sprawl, and subjective well-being. *Cities, 92*, 261–272. https://doi.org/10.1016/J.CITIES.2019.04.013

North, D. C. (2005). Institutions and the performance of economies over time. In *Handbook of New Institutional Economics* (pp 21–30).

Raadschelders, J. C. N. (1998). Evolution, institutional analysis and path dependency: An administrative-history perspective on fashionable approaches and concepts. *International Review of Administrative Sciences, 64*, 565–582. https://doi.org/10.1177/002085239806400403/ASSET/002085239806400403.FP.PNG_V03

Røsnes, A. E. (2021). Arealadministrasjon, 2nd edition. ed, ICB Research Reports. Universitetsforlaget, Oslo.

Schmidt, L. (2014). *Kompakt by, bokvalitet og sosial baerekraft*.

Skjeggedal, T., Nordtug, J., Wollan, G., & Ystad, D. (2003). Fortettingsrealisme. *Plan 35*, 56–63. https://doi.org/10.18261/ISSN1504-3045-2003-06-12

Statistic Norway. 2023. Table 09594: Land use and land cover (km^2), by area classification, contents, year and region and Table 11084: Tenure status. Households. Oslo.

Terje Holsen is an Associate Professor with the Department of Property and Law at the Norwegian University of Life Sciences (NMBU). He holds a Ph.D. in spatial planning and a M.Sc. in land consolidation (both NMBU). He also has urban design and comparative planning training from Oxford Brookes University. Between 2002 and 2008, he was Head of the Department of Landscape Architecture and Spatial Planning, NMBU, whereupon he was appointed to Director of

Estate at NMBU (2008–2015). His primary research interest lies in the interface between planning and real estate development, with a particular interest in institutional research on land policies.

Andreas Hengstermann is an Associate Professor with the Department of Urban and Regional Planning (BYREG) at the Norwegian University of Life Sciences (NMBU). He holds a Ph.D. in Geography from the Institute of Geography of the University of Bern (CH). He was educated as a planner (Dipl.-Ing. Raumplanung) at the TU Dortmund University (DE) and the Universidad de Huelva (ES). Furthermore, he did postgraduate studies in public law (DAS Law). His primary research interest lies in understanding the essential role of property rights in shaping spatial development. His research includes comparative studies across different geographical contexts, political systems, and legislations. He current serves as Vice-President International Academic Association on Planning, Law, and Property Rights.

Helén Elisabeth Elvestad is Head of Department and an Associate Professor with the Department of Property and Law (EIEJUSS) at the Norwegian University of Life Sciences (NMBU). She holds a Ph.D. in property law from NMBU and a master's degree in real property from NMBU. Her primary research interest are effects of land consolidation and the use of land readjustment as a measure in urban densification. Her research also includes property rights and planning law.

Land Policy in Poland: Evolution of the Liberalisation of Urban Planning and Policy Making

Małgorzata Barbara Havel and Tomasz Zaborowski

Abstract The post-communist period in Poland has been characterised by extensive liberalisation of spatial planning and a lack of coherent land policy. In recent years, the state has attempted to address problems arising from this liberalisation, but paradoxically, instead of strengthening the planning system and establishing a coherent land policy, it has further complicated the system by enacting various lex specialis—laws that take precedence over general planning regulations. Among these, the special Housing Act which aimed at facilitating housing development, has been particularly contested by the planning community. This act permits exceptions to established land use plans, but also imposes requirements on developers, particularly concerning the provision of urban infrastructure, which were previously non-existent. Additionally, the act has introduced the possibility of negotiations between municipalities and the developers. This chapter examines a case of successful application of the act, demonstrating its potential to foster housing development alongside the provision of public infrastructure. The case exemplifies the evolution of planning liberalisation and the reactive land policy at the municipal level.

1 Polish Understanding of "Land Policy"

In Poland, a policy that considers property rights and the public interest in spatial development is referred to as spatial policy (*polityka przestrzenna*). The term "land policy" (*polityka gruntowa* or *polityka terenowa*) is not formally established as a policy nor widely used in scientific or academic debates. It may refer to the creation and management of municipal land reserves, though this is perceived as a part of broader real estate management (*gospodarka nieruchomościami*) (Skalski, 2004). Nevertheless, this activity does not directly peruse the implementation of

M. B. Havel (✉)
Warsaw University of Technology, Warsaw, Poland
e-mail: malgorzata.havel@pw.edu.pl

T. Zaborowski
University of Warsaw, Warsaw, Poland
e-mail: t.zaborowski@uw.edu.pl

© The Author(s) 2025
T. Hartmann et al. (eds.), *Land Policies in Europe*,
https://doi.org/10.1007/978-3-031-83725-8_10

157

spatial policy goals. Recently, the implementation of urban planning objectives has been conceptualised as operational urbanism or operational urban development (*urbanistyka operacyjna*) (Ossowicz, 2019).

The absence of an officially formulated and explicit land policy does not imply that the public authorities do not engage in such practices. Several sectoral policies across different levels of governance and planning influence land allocation and distribution. In this chapter, we will trace one such implicit policy, referred to as the actual land policy. A case study of the transformation of an office business park in the city of Warsaw will illustrate how spatial policy shapes land development and influences the interaction between public planning and property rights in practice (Fig. 1).

It is argued that both the lack of official land policy and the actual land policy have roots in Poland's recent history. Following the fall of the communist regime in

Fig. 1 Służewiec in Warsaw, Poland (map: Kleiner, Jehling 2024, Aerial Imagery Provider © Esri, Maxar)

1989, the new state authorities deliberately chose to promote economic restructuring through market forces. The resulting liberal model of spatial development, characterised by a strong emphasis on private property, appears to be a natural consequence of that choice (Havel, 2022). Though dominant in Poland ever since, this approach has evolved over time. While maintaining a general laissez-faire stance, moderate turning points have occurred. For instance, the 2015 Act on Revitalisation introduced developer obligations through urban development contracts and measures to facilitate plan implementation. Meanwhile, the lack of a systemic approach was evident in the adoption of various *lex specialis*. From a land policy perspective, the most significant of these is the 2018 special Housing Act, which further liberalised the planning system by allowing deviations from land-use plans. This act may be viewed as an extension of planning liberalisation, however, it represents a more mature approach by introducing developer obligations as a condition for deviating from the provisions of land-use plans. This chapter reflects on the special Housing Act and presents a successful case of its application.

2 Understanding Polish Land Policy in Practice: Institutional and Policy Context for the Transformation of an Office Business Park in Warsaw

Warsaw, the capital and largest city in Poland is home to 1.97 million people, making it the sixth most populous capital city in the European Union. Situated in the east-central part of the country, Warsaw is also the capital of the Masovian Voivodeship (*Województwo mazowieckie*). In the Mokotów district, the Służewiec neighbourhood hosts one of the largest business office parks in Poland, occupying approximately 7.200 m2 opposite to one of the biggest shopping centres in the city (see Fig. 1). Originally a mono-functional office and commercial zone, Służewiec is currently experiencing a high demand for residential development, prompting the need to transform the Office Business Park area.

During the socialist era of the 1960s and 1970s, the Służewiec area was industrial, dominated by large factories and worker infrastructure. Following the collapse of socialist industries in the 1990s, the area became attractive to foreign real estate investors due to its prime location between the city centre and Chopin airport, the largest one in the country. The Office Business Park, developed on the former factory site, was the first major office complex in post-1989 Warsaw. With approximately 107,000 m^2 of office space, the park was built between 1996 and 2001, transforming the former industrial area into a modern business district.

During the rapid real estate boom of the 1990s, foreign investors and real estate experts spearheaded the transformation of this part of Warsaw by leveraging their global knowledge and international networks (Havel, 2022). Today, Warsaw has a mature real estate market with nearly 5.9 million m^2 of office space (Savills, 2021) with over 90 per cent of commercial real estate investment coming from foreign entities. Next to the Office Business Park, the area has transformed into a dense office cluster, attracting a large influx of daily white-collar commuters. However, as the local road systems were not designed for such heavy traffic, this has led to severe congestion and parking shortages. The whole neighbourhood is colloquially referred to as "Mordor", a reference to the evil land in the fantasy novel *The Lord of the Rings* by J.R.R. Tolkien. This nickname is so widely recognised that the area has such Facebook fan page with over 170,000 followers, and two streets are officially named Tolkien and Gandalf.

To address the district's lack of functional diversity, developers have begun adding residential buildings to balance the supply between housing and offices. Today, Służewiec is being transformed into a mixed-use area, with plans for offices, apartments, parks, and a school. The following section will detail how this transformation is being managed through land policy and planning instruments.

2.1 Planning Permissions in the Polish System

In Poland, building permits can be obtained via four primary pathways, one of which was used for the transformation of the Office Business Park.

1. Binding Detailed Land-Use Plans (MPZP): The first option is binding detailed land-use plans (MPZP, from the Polish *miejscowy plan zagospodarowania przestrzennego*). The MPZP is the traditional method of urban planning, governing land use at the local level. These plans are developed by municipalities based on a higher-level land-use strategy, called the Study of Conditions and Directions for Spatial Development as well as planning law.
2. Decisions on Land Development Conditions (DWZ): The second option is based on local administrative decisions, called the decision on land development conditions (the DWZ, from the Polish *decyzja o warunkach zabudowy*). This is the dominant pathway for development in areas without a binding land-use plan. The DWZ operates on the 'good surroundings' principle, which allows development if specific conditions regarding land use and infrastructure are met: (i) at least one neighbouring plot accessible from the same public road is developed in a way that aligns with the proposed land use, density, and architectural form; (ii) the land has access to a public road; (iii) existing or planned infrastructure is sufficient; (iv) no permission is required to change the land's use from agricultural or forestry purposes; and (v) the development complies with other specific regulations (e.g. the Act on Environmental Protection) (Art. 61.1 LUDPA). The DWZ has long been criticised for its vagueness, which has allowed for virtually unrestricted development. Only in 2022 was the scope of the surrounding area to be analysed more clearly defined, though significant discretion remains with authorities.
   Similar regulations exist in other European countries, such as Belgium and Germany (Hengstermann & Hartmann, 2021). However, the DWZ differs in that

it enables development outside densely built-up areas, which has contributed to extensive suburbanisation and peri-urbanisation in rural areas.

3. National Special Acts: National special acts (from the Polish *specustawy*) offer a third option for development, significantly altering the spatial planning system. These laws bypass other regulations, such as those governing spatial planning, land management, and construction. Special acts streamline multiple procedures, including obtaining local administrative decisions, building permits, and decisions on property division or expropriation. Currently, nineteen special acts are in effect, covering areas such as railroads, roads, the Central Transportation Port, housing, an electric car factory (Majda & Havel, 2022).

4. Local Revitalisation Plans (MPR): The fourth option, the Act on Revitalisation of 2015 (*Ustawa o rewitalizacji*) introduces specialised land-use regulations and instruments for revitalisation areas. Municipal councils may adopt a local revitalisation plan (MPR, from the Polish *miejscowy plan rewitalizacji*), to outline the responsibilities of both the investors and the municipality, such as building technical and social infrastructure and residential premises. This can be formalised through a town planning contract (*umowa urbanistyczna*), although (Havel & Załęczna, 2022).

In the case of the office complex, a binding land-use plan (MPZP) adopted in 2011 designated the area for commercial and administrative services use only, preventing residential development. However, the site is economically very attractive, with a lot of unused space and potential for development. Conflicts arose due to the MPZP's focus solely on office construction, ignoring housing demand. Until 2018, it was impossible to change the land use of the site without amending Warsaw's overarching preparatory land-use plan. This situation changed with the introduction of the special Housing Act (from the Polish *specustawa mieszkaniowa*[1]), established as a part of the National Housing Program, which ultimately enabled the site's recent transformation into a residential area.

2.2 Contradictions, Crisis, and Housing Needs

This case study highlights the major tensions in Poland's land use policies. Post-socialist cities have changed beyond recognition, and while many positive changes are evident—particularly in the development of the commercial real estate market, exemplified by the aforementioned growth of office parks—the Polish urban landscape often diverges from official planning authority goals. Despite successes in urban projects and revitalisation efforts, the country continues to lean towards a neoliberal, highly liberalised approach to planning.

---

[1] Act on Facilitating the Preparation and Implementation of Housing and Associated Facilities (Journal of Laws of 2018, item 1496)—Ustawa z dnia 5 lipca 2018 r. o ułatwieniach w przygotowaniu i realizacji inwestycji mieszkaniowych oraz inwestycji towarzyszących (Dz. U. 2018 poz. 1496).

Currently, only 31.4% of Poland's land is covered by legally binding local development plans (MPZP) (CSO, 2021), while the whole of the country is covered by higher-level preparatory land-use plans which are only binding for the MPZP. However, land designated for housing has been overestimated in these plans by five or six times (Kowalewski et al., 2014). Compounding this issue is the fact that decisions on land development conditions (DWZ) do not have to adhere these preparatory plans, making it difficult to estimate the amount of actual developable land (Zaborowski, 2021). Despite the oversupply of building land, housing needs remain unmet, while the rigidity of the MPZP stifles development, as in the case of the office park.

The majority of urban development in Poland is driven by local administrative decisions, the DWZ. According to a 2007 report by the Supreme Audit Office (Najwyższa Izba Kontroli, or NIK), 72per cent of building permits were granted through administrative decisions with only 28 per cent based on MPZP (Raport Najwyższej Izby Kontroli NIK, 2007). By 2011, 68 per cent of building permits (170,00 out of 250,000) were issued through DWZs (Majda & Havel, 2022). This trend has significantly contributed to the prevalence of greenfield developments, which constituted 58 per cent of total capital investment in 2004 (Majda & Havel, 2022). DWZs also serve as a tool to convert agricultural land to building plots, further expanding development in areas not intended for urbanisation. In 2022, a NIK report revealed that 36 per cent of DWZs were issued in areas not intended for development under higher-level preparatory plans.

This unregulated expansion has resulted in widespread functional and visual chaos, particularly in the suburbs of major cities, contributing to their haphazard suburbanisation (Solarek, 2013). These developments undermine the goals of sustainable urban planning (Zaborowski, 2014). Poland's Main Urban Planning and Architectural Commission acknowledged this crisis in 2011, citing the lack of spatial coordination of development processes as a major obstacle to the proper functioning of urban and rural settlements. This chaotic growth has worsened living conditions and diminished the social, cultural, and economic value of spaces, while the unchecked, neoliberal "freedom of development" has deepened social and financial inequalities due to insufficient urban infrastructure (Havel, 2022). The financial burden of this spatial disorder is estimated at 84.3 billion zlotys per year[2] (Kowalewski et al., 2014).

The situation in Poland raises the question of whether the country's actual land policy contradicts its official goal of sustainable spatial development, or, if the problem lies in the absence of an effective land policy to counteract spontaneous and uncoordinated growth. The roots of this issue can be traced back to the post-communist era where property rights underwent a dramatic shift (Havel, 2020). After the fall of communism, which had emphasised public ownership, a reverse approach emerged—an 'absolutisation' of private property (Jędraszko, 2005, p. 73). This liberal approach, often referred to as 'holy property rights' (*święte prawo własności*), has led to an equivalence between property rights and development rights (Izdebski, 2013, pp.151–154). The so-called 'freedom of development' (*wolność budowlana*) has since characterised Polish land policy (Koziński, 2012; Niewiadomski, 2009),

---

[2] Approximately 19,5 billion euro per year.

resulting in a highly liberalised approach to shaping the landscape (Zaborowski, 2021).

One consequence of this approach is the high cost associated with enacting detailed land-use plans, as property owners are entitled to compensation for any reduction in land value caused by planning regulations. This creates a significant imbalance in favour of private developers and landowners (Havel, 2017), with limited use of public value capture (PVC) instruments (Gdesz, 2012; Havel, 2017, 2020). Meanwhile, Poland faces a severe housing shortage, particularly affecting middle- and low-income households. The housing deficit is estimated at 641,000 dwellings (about 4.5 per cent of households) (National Housing Programme, 2016), with current housing conditions falling below the European Union member state average. With 363 flats per 1,000 people, compared to the EU average of 435, at least 500,000 new affordable homes are needed in Poland, yet only a few thousand are built annually (Habitat for Humanity Poland, 2022). The progressive commodification of housing units has further exacerbated the situation, making units an investment vehicle for the wealthy, while the privatisation of housing stock has led to nearly 90 per cent of flats being privately owned (Sagan, 2021).

To address the housing crisis, the Law and Justice government adopted the National Housing Programme (a.k.a. Apartment Plus Programme) in 2016. This strategic plan, aimed at reducing the housing deficit by 2030, seeks to improve housing conditions through a combination of legal, financial, and organisational reforms. Unlike previous policies, the program includes measures such as rental market rules, housing cooperatives, affordable housing, and improved land development and construction processes. A key legislative development was the introduction of the Act on Facilitating the Preparation and Implementation of Housing and Associated Facilities (the Housing Act). This temporary *lex specialis* was intended to streamline housing delivery for moderate-income groups by reducing administrative and legal barriers. While the Act has drawn criticism for promoting development that bypasses existing land-use plans, it also introduces value capture instruments and allows for greater coordination between public and private parties. The Act represents a reactive land policy aimed at facilitating housing delivery by relaxing rigid planning restraints. It further introduces rules regarding infrastructure costs and urban infrastructure availability, integrating these considerations into Polish planning legislation. This approach, as applied to the office park redevelopment, highlights the Act's potential to balance private development interests with broader urban planning goals.

3   Actors in Polish Land Policy

While numerous actors are involved in land policy in Poland, the main private and public actors relevant to the redevelopment of the Office Business Park are outlined below.

3.1 Private Actors and Property Ownership

The Office Business Park in *Służewiec* has been divided and it is now owned by two private landowners, both commercial real estate companies. The newer buildings, constructed in the late 1990s and early 2000s, are owned by a Vienna-based company. Meanwhile, the land with refurbished buildings was sold in 2021 to a Polish developer with extensive experience in the main sectors of the real estate market, particularly in transforming entire neighbourhoods into multifunctional urban areas.

In Poland, private actors hold significant influence in urban development, as ordinary public actors typically have limited power to enforce active land policies. In this context, the most powerful private actors are property owners and developers. In spatial planning, private property is considered equally important to the public interest. The office park land use can be influenced by different land policy instruments. Although municipalities decide which land policy instruments will be used to change land use, it is ultimately up to the developer to decide when to start the redevelopment.

Owning any kind of real estate is very popular in Poland. A key objective of the state's current land policy is to encourage private home ownership. Most state housing policies introduced in recent years have focused on fostering the construction or purchase of private homes or flats. Given the widespread ownership of land, there is a strong opposition to any limitations on land development rights.

Both landowners and developers benefit from the liberal property market and weak land policy. The ineffectiveness of public value capture (PVC) tools increases their profits from urban development. This dynamic has made Poland a haven for developers and landowners, often at the expense of public urban infrastructure quantity and quality. Institutional developers can expect higher profits compared to investments in countries with effective PVC tools.

3.2 Public Actors and Institutions

To understand which public authorities influence land use in the Office Business Park area, it is important to examine the way land policy is implemented in Poland.

Land policy in Poland operates at two administrative levels: national and municipal. At the national level, there is currently no comprehensive plan nor spatial planning policy guidance. The only national document related to urban development is the National Urban Policy 2030 (*Krajowa Polityka Miejska* 2030). While this document addresses key challenges, such as suburbanisation and the shortage of affordable housing, it does not establish a formal land policy and lacks implementation power.

Despite the absence of a formal land policy, the Polish state engages in piecemeal land policies through extraordinary enactments (*specustawy*) to address urgent needs like the effective construction of roads, railways, and airports, or housing shortage. These acts circumvent the rigid spatial planning system by introducing special rules that allow for stronger interventions in property rights, such as integrating administrative permissions (DWZ) with land subdivision in a single decision or by acquiring land by law rather than through the longer expropriation process typically required of public purposes.

The most significant administrative level for land policy is the municipality (*gmina*) that has s.c. planning independence. The mayor is responsible for preparing detailed land-use plans and issuing decisions on land development conditions (DWZ), but it is the municipal council that approves these plans through a resolution. The municipal council also has the authority to approve or reject a developer's application for housing development under the housing *lex specialis*.

According to the Real Estate Management Act (*ustawa o gospodarce nieruchomościami*), municipalities are expected to create land stock, but strategic land banking is rarely practiced in Poland. Typically, municipalities acquire land only for public purposes, such as road construction, and more often sell land properties than purchase them. Deliberate municipal land policy, a strategic component of land management, is generally lacking. However, in recent years, some big cities like Wrocław have attempted to implement s.c. operational urban development (*urbanistyka operacyjna*) measures, including institutionalised negotiations with developers to co-finance public urban infrastructure.

3.3 Reactive Land Policy *via* Special Housing Act

The strategies of land policy can be classified as active, passive, reactive, and protective land policy (Shahab et al., 2021). The case of the Office Business Park exemplifies a reactive land policy which is demand-oriented and responds to market needs. In this case, the developer seeks to change the development's character to residential use, contrary to the provisions of the local land-use plan (MPZP). What opportunities does this situation present?

Under Polish law, landowners cannot directly approach municipalities to negotiate additional or different building rights for an existing land-use plan or sign an urban development contract. Urban planning and development agreements do not exist as planning instruments in the context of land-use plans (MPZP). Prior to 2015, Polish legislation did not allow for negotiations between private and public parties in land development. The only exception was the construction of public roads during the application process for a building permit—typically the final stage of development (Havel & Załęczna, 2022; Munoz Gielen et al., 2022; Zaborowski, 2018). In planning law, plan preparation and plan implementation processes are distinct processes: local authorities prepare physical plans without developer involvement. Although the developer can request the creation of a new plan (MPZP), it must align with the higher level preparatory land-use plan.

In 2015, the Act on Revitalisation (*Ustawa o rewitalizacji*) introduced a new instrument—a town planning contract (*umowa urbanistyczna*) for revitalisation areas. In other areas, new opportunities for negotiations between private and public parties only emerged in 2018 with the introduction of the special Housing Act. This legislation aimed to reduce the administrative and legal barriers to land acquisition for housing development.

Under the Housing Act, significant residential developments may be permitted even if they do not align with the existing MPZP, provided they do not contradict upper-level preparatory land-use plans (the Studies). However, this requirement does not apply to post-industrial, post-military, post-railway and post-postal areas, meaning new housing developments on these sites do not have to comply with any plans.

The Housing Act stipulated that a housing project should involve the (re)construction of multi-family residential buildings with at least 25 residential units or 10 single-family residential buildings. Associated facilities must include access to a public road, public transport infrastructure, cultural activities, childcare, nurseries, schools, daycare support, healthcare, sports and recreation, and green areas serving the development's residents. Developers wishing to start a housing project must submit an application to determine the project's location (*Wniosek o ustalenie lokalizacji inwestycji mieszkaniowej i inwestycji towarzyszących*) to the municipal council and begin negotiations with the municipality. The application must include many legally required documents, such as an urban and architectural concept of the project and assurances of access to required technical and social infrastructure. The special Housing Act respects the principle that the planning authority rests with the municipal council and its residents. The council retains the right to refuse any housing project that conflicts with the municipality's needs and development potential.

The Act establishes several conditions for starting a housing project, including direct access to a public road, connections to water and sewage systems, and access to the electricity grid. It also introduced new requirements for maximum distances to schools, kindergartens, and recreational areas, known as urban standards. A significant change to the planning system is that developer may sign an agreement with the municipality (*porozumienie z inwestorem*) to provide the necessary technical and social infrastructure if it, lacking. This agreement forms the basis for meeting urban standards.

Thus, the Act encourages developers to take on the cost of urban infrastructure and facilitates negotiations between landowners and local authorities (representing residents) over these costs. By applying urban standards, the Act also aims to achieve spatial quality goals in urban development. According to the legislator, these standards are intended to curb uncontrolled urban sprawl. In Poland, new housing estates are often located on city outskirts, based on land development conditions (DWZ) that in fact do not require developers to provide urban infrastructure. Local authorities can adjust the default maximum distances in urban standards by up to 50%. The municipal council can do this by enacting local urban standards.

During the legislative process, the new Housing Act sparked significant controversy and was dubbed the "Lex Developer Act" by the political opposition. The new legal framework was imposed on municipalities without pilot studies. According to Warsaw's authorities, adopting local urban standards was crucial to mitigating the negative effects of the Housing Act. Warsaw was quick to act, with its council adopting local urban planning standards on 30 August, 2018.

In the example of the Office Business Park, a landowner applied for a residential project involving the construction of multi-family residential buildings (7,000 m², 1,250–1,600 housing units) with associated public infrastructure. The developer was committed to providing public green space with landscaping, sidewalks, small architectural elements, leisure, recreational, and sports facilities, necessary technical infrastructure, parking lots, and bicycle paths. Additionally, the developer agreed to build and transfer ownership of a public elementary school, complete with all necessary infrastructure, to the city.

In their application, the developer justified the need to transform this part of the city, by referencing existing urban and strategic planning documents adopted by the City of Warsaw. They highlighted the shortage of space for multi-family residential areas, the need to create neighbourhoods equipped with diverse infrastructure and common spaces, and the goal of improving the public space quality and enhancing the city's polycentric structure.

The Warsaw Council reviewed the developer's proposal twice. Initially, the council denied permission. The developer then returned with a revised proposal, increasing their investments in public developments, including fully financing the construction of a three-story elementary school on their other land. In doing so, this revision met the expectations of the Mokotów authorities, who wanted a larger school. Approval was granted in 2021, making it the largest development in Warsaw under the special Housing Act. Demolition of the old office buildings is currently underway.

4 Reflection

The post-communist period in Poland has been characterised by extensive liberalisation of spatial planning. The case of the Office Business Park illustrates the ongoing evolution of this approach. On the one hand, it further highlights liberalisation by allowing developers to bypass certain land-use regulations. On the other hand, it demonstrates the increasing importance of providing urban infrastructure in development requirements and public discourse.

The special Housing Act, often pejoratively referred to as "Lex Developer", faced significant criticism from the planning community. Initially, the professional debate overlooked a key aspect of that Act: its potential to link urban development with the previously neglected provision of public infrastructure, the cost of which now could be transferred to the developer. Additionally, the Act's facilitation of official negotiations between municipalities and developers represented a crucial advancement. This aligns with global trends towards greater flexibility in urban development process, signalling a significant shift in Poland's planning culture. This shift may hopefully pave the way for a more cooperative approach to planning and urban development.

Under the special Housing Act, developers became responsible for meeting infrastructure standards. The act granted municipalities a powerful tool to refuse permits if developers were unwilling to contribute to financing public urban infrastructure. This example was vividly illustrated when the city of Warsaw rejected the first application for the redevelopment of Office Business Park. It was only after this rejection that the developer agreed to meet the city's financial demands in order to secure approval from the City Council.

While the legal framework did not permit negotiation over the adoption of a land-use plan (as, for example, in Germany), it did allow for negotiation of deviations from existing regulations. In practice, the outcome of these two approaches is essentially the same: housing supply increases, and urban infrastructure is largely provided at no cost to the municipality. The municipality retains control over land use, albeit in a reactive manner.

As of the writing this chapter, the President of the Republic of Poland has signed an amendment to the planning law that introduces several significant changes to the planning system. One of these changes is the abolition the special Housing Act, effective on 1st January 2026. It will be replaced by a new instrument called the integrated development plan (*zintegrowany plan inwestycyjny, ZPI*). This new legal framework will allow the adoption of such plans to be negotiated, thereby maintaining the flexibility that characterised the special Housing Act. Crucially, the shift towards a cooperative regime has now been enshrined in the main Planning Act.

Could the development scheme under the new law be implemented? Yes. Rather than deviating from its current land-use plan, a developer would apply for a new ZPI, and the municipality could negotiate approval in exchange for a similar set of contributions.

The case presented may not be representative of the entire country. It is the largest scheme of this kind in Poland and involves the most significant developer obligations.

Overall, the special Housing Act has not met its initial expectations. On the one hand, it has had a marginal impact on the housing market.[3] Research by Załęczna and Antczak-Stępniak (2022) shows that legal framework provided by the Housing Act regulations has rarely been chosen by developers. As of May 2022, about 200 local resolutions of this kind had been issued (Załęczna & Antczak-Stępniak, 2022, p. 89). On the other hand, the act did not significantly increase developers' contribution to public urban infrastructure, with 71 per cent of municipal resolutions concerning housing developments not including accompanying infrastructure. In these cases, the existing infrastructure was deemed sufficient. In only 12 out of 30 cases investigated by Załęczna and Antczak-Stępniak (2022) did developers commit to providing new social infrastructure, and in most instances, this was limited to children's playgrounds.

In conclusion, the critics of the special Housing Act were largely correct: developers tended to exploit the act without significantly contributing to public amenities. However, the presented case also illustrates the act's potential to enhance the quality of urban development and marks a noteworthy stage in the evolution of Poland's planning system.

---

[3] There are ca. 450 applications yearly (Monitoring, 2023).

References

CSO. (2021). Central statistical office of Poland. Retrieved January 19, 2022, from https://stat.gov. pl/en

Gdesz, M. (2012). Rozkładanie kosztów urbanizacji. In I. Zachariasz (Ed.), *Kierunki reformy prawa planowania i zagospodarowania przestrzennego*. Wolters Kluwer Polska

Habitat For Humanity Poland. (2022). Dlaczego pomagamy. Retrieved January 19, 2022, from https://habitat.pl/dlaczego-pomagamy/

Havel, M. B. (2017). How the distribution of rights and liabilities in relation to betterment and compensation links with planning and the nature of property rights: Reflections on the polish experience. *Land Use Policy, 67*, 508–516. https://doi.org/10.1016/j.landusepol.2017.06.032

Havel, M. B. (2020). The effect of formal property rights regime on urban development and planning methods in the context of post-socialist transformation—An institutional approach. In R. Levine-Schnur (Eds.), *Measuring the effectiveness of real estate regulation* (pp. 149–169). Springer. https://doi.org/10.1007/978-3-030-35622-4_8

Havel, M. B. (2022). Neoliberalization of urban policy-making and planning in post-socialist Poland—A distinctive path from the perspective of varieties of capitalism. *Cities, 127*, 103766. https://doi.org/10.1016/j.cities.2022.103766

Havel, M. B., & Załęczna, M. (2022). Poland. In: J-M. Halleux, A. Hendricks, B. Nordahl, & V. Maliene (Eds.), Public value capture of increasing property values across Europe (pp. 183–192). vdf Hochschulverlag.

Hengstermann, A., & Hartmann, T. (2021). Grund zum Wohnen—Das Baulandmobilisierungsgesetz aus internationaler Perspektive. *PND—Rethinking Planning, 1*(30–41). https://doi.org/10. 18154/RWTH-2021-01677

Izdebski, H. (2013). Ideologia i zagospodarowanie przestrzeni: Doktrynalne prawno-polityczne uwarunkowania urbanistyki i architektury. Wolters Kluwer Polska.

Jędraszko, A. (2005). *Zagospodarowanie Przestrzenne w Polsce—Drogi i Bezdroża Regulacji Ustawowych*. Unia Metropolii Polskich.

Kowalewski, A., Mordasiewicz, J., Osiatyński, J., Regulski, J., Stępień, J., & Śleszyński, P. (2014). Ekonomiczne straty i społeczne koszty niekontrolowanej urbanizacji w Polsce—Wybrane fragmenty raportu. *Samorząd Terytorialny, 1*, 5 21.

Koziński, J. (2012). Doktryna swobody budowlanej. Aspekty ekonomiczne i urbanistyczne. In: Problemy Planistyczne—Wiosna 2012. *Zeszyt Zachodniej Okręgowej Izby Urbanistów, 1*, 5–24.

Majda, T., & Havel, M. B. (2022). Po wejściu do Unii Europejskiej (2004–2022). In: J. Korzeń (Ed.), *Stulecie towarzystwa urbanistów polskich 1923–2023* (pp. 177–214). Narodowy Instytut Architektury i Urbanistyki.

Monitoring. (2023). Ministerstwo Rozwoju, Pracy i Technologii. Monitoring ustawy z dnia 5 lipca 2018 r. o ułatwieniach w przygotowaniu i realizacji inwestycji mieszkaniowych oraz inwestycji towarzyszących.

Munoz Gielen, D., Ossowicz, T., & Zaborowski, T. (2022). Failure and opportunities of public value capture and developer obligations in Polish urban development. *Miscellanea Geographica, 26*(1), 15–30. https://doi.org/10.2478/mgrsd-2020-0071

National Housing Programme (2016). Narodowy Program Mieszkaniowy, uchwała nr 115/2016 Rady Ministrów z dnia 27 września 2016 r.

Niewiadomski, Z. (Ed.) (2009). Prawna regulacja procesu inwestycyjno-budowlanego. Uwarunkowania, bariery, perspektywy. Perspektywy. LexisNexis.

Ossowicz, T. (2019). Urbanistyka operacyjna. Zarys teorii. Oficyna Wydawnicza Politechniki Wrocławskiej.

Raport Najwyższej Izby Kontroli NIK (2007). Informacja o wynikach kontroli kształtowania polityki przestrzennej w gminach jako podstawowego instrumentu rozwoju inwestycji. Warszawa 2007.

Sagan, I. (2021). Polityka miejska w warunkach kryzysu. In: A. Nowak (Ed.), Polityka przestrzenna w czasie kryzysu. Wydawnictwo Naukowe Scholar.

Savills (2021). Wiadomości Savills Retrieved January 5, 2022, from https://www.savills.pl/wia domosci-i-raporty/savills-wiadomosci/310305/ponad-5-mld-euro-zainwestowano-w-polsce- w-ni eruchomo%EF%BF%BD%EF%BF%BDci-komercyjne-w-2020-roku–z-czego-połow ę-w-magazyny.

Shahab, S., Hartmann, T., & Jonkman, A. (2021). Strategies of municipal land policies: Housing development in Germany, Belgium, and Netherlands. *European Planning Studies, 29*(6), 1132– 1150. https://doi.org/10.1080/09654313.2020.1817867

Skalski, K. (2004). Rewitalizacja a instrumenty zarządzania przestrzenią miast polskich. *Problemy Rozwoju Miast, 1*(3–4).

Solarek, K. (2013). Struktura przestrzenna strefy podmiejskiej Warszawy. Oficyna Wydawnicza Politechniki Warszawskiej.

Zaborowski, T. (2014). Suburbanization in the light of sustainable spatial development principles. In M. Czerny, & G. Hoyos Castillo, (Eds.), *Suburbanization Versus Peripheral Sustainability of Rural-Urban Areas Fringes* (pp. 1–38). Nova Science Publishers.

Zaborowski, T. (2018). Land acquisition and land value capture instruments as determinants of public urban infrastructure provision: A comparison of the Polish legal framework with its German counterpart. *Geographia Polonica, 91*(3), 353–369.

Zaborowski, T. (2021). It's All about Details. Why the Polish Land Policy Framework Fails to Manage Designation of Developable Land. *Land*, 10, 890. https://doi.org/10.3390/land10 090890

Załęczna, M., & Antczak-Stępniak, A. (2022). „Lex Developer" in practice—The scale of application in the largest polish cities. *Real Estate Management and Valuation, 30*(4), 86–97.

Acts of Legislation

The Act on Facilitating the Preparation and Implementation of Housing and Associated Facili- ties, (Ustawa z dnia 5 lipca 2018 r. o ułatwieniach w przygotowaniu i realizacji inwestycji mieszkaniowych oraz inwestycji towarzyszących (Dz.U. 2018 poz. 1496)).

The Act on Revitalization of 2015 (Ustawa z dnia 9 października 2015 r. o rewitalizacji (Dz.U. 2015 poz. 1777)).

The Land Use Planning and Development Act of 2003 (LUPDA) (Ustawa z dnia 27 marca 2003 r. o planowaniu i zagospodarowaniu przestrzennym (Dz. U. 2003 Nr 80 poz. 717)).

Małgorzata Barbara Havel is an urban planner and designer, researcher of spatial planning systems and land policy instruments in Europe. She is an assistant professor in the Chair of Spatial Planning and Environmental Sciences at the Faculty of Geodesy and Cartography, Warsaw Univer- sity of Technology. After studying at the Faculty of Architecture at the Warsaw University of Technology, she earned her doctorate in the research field of real estate economics at Helsinki University of Technology (now Aalto University). From 2011–13 she worked at the University of Cambridge, Department of Land Economy, and Department of Architecture, and from 2014– 17 at the Norwegian University of Life Sciences. She is a member of the Society of Polish Town Planners.

Tomasz Zaborowski is an assistant professor in the Chair of Urban Geography and Spatial Plan- ning at the Faculty of Geography and Regional Studies, University of Warsaw. He graduated from Wrocław University of Technology where he obtained a master's degree in architecture and urban planning (2004) and doctor's degree in spatial planning (2013). He is a member of the Task Force for Legal and Urban Planning at the Committee for Spatial Economy and Regional Planning, Polish Academy of Sciences. His scientific interests range from sustainable urban development and land policy instruments to comparative research of legal planning frameworks. He studied, taught or investigated in Poland, Germany, England, Spain and Colombia.

Land Policy in Sweden: The Paradoxical Difficulty of Creating Affordable Housing at the Intersection of Municipal Land Allocation, Local Planning and Housing Policy

Anna Granath Hansson

Abstract Based on a long tradition of municipal land purchases, today 70–80 percent of housing development in Stockholm takes place on land allocated by the city. To densify the urban core, several larger developments are being realised on brownfield land. Norra Djurgårdsstaden is one of the most attractive development areas and ambitions regarding environmental sustainability and design are high. The Stockholm municipal plan states that affordable housing and social mix are important public goals. Land policy is mentioned as one of the main instruments to realise these goals. Simultaneously, Swedish housing policy emphasises 'good housing for all' and does not prioritise lower income households. Affordable housing is defined as rental housing irrespective of rent level. Recently, critic has arisen that Norra Djurgårdsstaden is only accessible to higher income households in contrast to city policy goals. This chapter attempts to explain incentives of the city to avoid fulfilling its own policy goals, linking this paradox to planning and land policy on the one hand and housing and fiscal policy on the other hand, showing how varied public interests are difficult to balance. Several other Swedish municipalities face the same difficulties in using land policy to meet affordable housing and social mix goals.

1 Land Policy in Sweden

Land policy in Sweden is captured within a number of policies related mainly to planning, building, land readjustment and environmental protection. Related laws and regulations, and their mode of implementation, regulate the balance between private property rights and the public interest. Environmental sustainability is much emphasised and often takes precedence in planning and city development. The term land policy translates into *markpolitik* in Swedish and is usually used in relation to municipal attempts to advance housing development through planning and land

A. Granath Hansson (✉)
Nordregio, Stockholm, Sweden
e-mail: anna.granath.hansson@nordregio.org

© The Author(s) 2025
T. Hartmann et al. (eds.), *Land Policies in Europe*,
https://doi.org/10.1007/978-3-031-83725-8_11

allocation. When combined, planning and land allocation are powerful municipal tools to steer city development. The balance between state, municipal and developer interests has been much debated over a longer time. According to the law on municipal land allocation, which was introduced in 2014, municipalities are obliged by the state to have guidelines that describes priorities, procedures and price-setting principles related to municipal land allocation to increase transparency.

Due to strong municipal self-government rights and municipal planning monopoly, the state and regions only have very limited decision rights on the local level. However, the county administrative boards *(länsstyrelserna)*, that represent the state in the 21 counties, have the right to cancel local development plans in case they are not compatible with five specific state interests (Kalbro & Lindgren, 2021). Such interests are inter alia related to the environment, infrastructure, industries, the health and security of the population and coordination between municipalities. Although regions *(regioner or landsting)* have stepped up their ambitions in regional planning expanding also into housing related issues in recent years, it should be noted that regional plans do not have a strong position in the Swedish setting (for example Smas & Schmitt, 2021). However, in Stockholm, regional planning coordinating the 26 municipalities in the region is obligatory according to the planning and building code *(Plan- och bygglagen)*.

The real power in land policy in Sweden rests with the 290 municipalities. The planning and building code stipulates already in Chap. 1, paragraph 2 that 'It is a municipal matter to plan the use of land and water according to this law'. Such planning includes an overall municipal plan, area regulations and local development plans. Only the local development plans are binding towards third parties. Many municipalities have local development plans for most of its built-up area, and thus in a sense pursue a proactive land policy. When areas with development potential are (re)planned, this is often done in negotiation with developer(s). However, this procedure cannot be labelled reactive as many municipalities use their planning monopoly to advance municipal interests.

Further, many Swedish municipalities own substantial parts of the land within their jurisdictions. The City of Stockholm has pursued an active land banking policy for more than 150 years (Sheiban, 2002). The aim of such purchases has mainly been to steer city development and facilitate development of affordable housing (Caesar, 2016; Sidenbladh, 1981). Today, the city owns an estimated 70–80 percent of the land planned for housing development. The dual ability to steer through land ownership and planning provides the city with a decisive influence on city development. *Norra Djurgårdsstaden* (NDS), the city-owned development area that is presented in this text, is an example of how this active land policy is pursued. The case is illustrative for Swedish land, planning and housing policy, as many municipalities face the same difficulties as in NDS when developing their land.

2 Why is There no Affordable Housing in Norra Djurgårdsstaden?

As many other cities, Stockholm aims to densify the city by developing land that is considered underused. Several larger city developments have been or are being realised on brownfield sites. The NDS development area is the grounds of a former gasworks and a harbour. The area combines proximity to the most attractive central areas of the capital, with all that it entails of public and private services, with prompt access to nature as it borders Sweden's only city national park and the sea (see Figs. 1 and 2). Administratively, the area belongs to the Östermalm city district, which is one of the most expensive housing areas in the country. In this development area, the city has very high ambitions when it comes to landscape and building design, as well as environmental sustainability (Stockholm, 2021).

The former industrial zone is currently being transformed into a mixed-use area dominated by housing, but also containing shops, restaurants, offices, schools and daycare, museums, leisure venues, as well as playgrounds and parks. Some of the architecturally valuable gasworks buildings are being transformed into public space such as a concert hall, a museum, shopping facilities and restaurants. Planning of the area began in the early 2000s and the first phase completed in 2012. When fully built out the area is expected to house 12,000 inhabitants and provide 35,000 working places (Stockholm, 2021). The development area is owned by the City of Stockholm. Development is steered through municipal land ownership and the city's planning monopoly. The area is a city environmental show case where high demands on environmental sustainability include greenery and urban gardening, mobility, handling of waste etcetera. Land prices reflect the attractivity of the area and do not allow for housing that moderate- and lower-income households could afford. Such housing is accordingly not required by the city despite explicit housing policy goals related to housing provision 'for all' and social mix.

As the project has evolved, critic has arisen that the area is only accessible to higher income households in contrast to city policy goals. Critics among politicians and housing interest organisations claim that as the city has the planning monopoly and owns the land, there should be room to steer towards affordable housing should there be a wish to do so. However, so far, there has been a reluctance to introduce affordable housing schemes as it is deemed politically sensitive, complicated and potentially inefficient. Potential tools to introduce housing that would be affordable to moderate-income groups in otherwise market-rate housing areas are tested on a smaller scale in less attractive parts of the city.

This chapter attempts to explain incentives of the city to avoid fulfilling its own policy goals, linking this paradox to planning and land use on the one hand and housing and fiscal policy on the other hand, showing how the municipality has triple roles as authority, landowner and developer which are difficult to balance. Stockholm's choice not to pursue its own policy for affordable housing and social mix in most new housing developments can be studied in a multitude of Swedish municipalities. Attempts to introduce such policy have had little success, based on similar

Fig. 1 Norra Djurgårdsstaden in Stockholm, Sweden (map: Kleiner, Jehling 2024, Aerial Imagery Provider © Esri, Maxar, Microsoft)

situations and use of policy tools as in Stockholm (for example Granath Hansson et al., 2024).

3 The Cooperation Between the Municipality and Public and Private Property Developers

The city's vision for the housing market is 'good opportunities for all to live in good housing' (Stockholm, 2018, p. 45). In the municipal plan, housing provision is pointed out as the most important public interest, which shall guide municipal planning. Affordable housing is generally understood as rental housing built both by municipal and private housing companies. In attractive locations, construction of

Fig. 2 The stretch of buildings along the cove Husarviken with two of the listed gasometers in the background. Photo by Jansin & Hammarling

rental housing is also seen as the main method to create neighbourhood social mix (Caesar & Kopsch, 2018). According to the Stockholm Housing Provision Program, extending the affordable housing supply and working towards social mix are top priorities (Stockholm, 2020). In the budget for 2024 this goal is further underlined, also emphasising the role of municipal housing companies and lower cost serial housing concepts, including the city initiative Stockholmshusen (Stockholm, 2023). The large land bank of the city, the city's responsibility for physical planning and its three municipal housing companies are mentioned as important tools to implement the policy (Stockholm, 2018). The three municipal housing companies are of a good financial standing and have the goal to build 1500 apartments a year. An unspecified share of new development is intended to be rental housing that is affordable to the young, students and 'other groups with a weak position on the housing market' (Stockholm, 2020, p. 44).

The city owns the larger share of the land that is planned for housing (Stockholm, 2020). When land banking started in the nineteenth century, land purchases aimed at steering city development and creating better conditions for new affordable housing (Sheiban, 2002). Today, the steering component is the main driving force in continued land acquisition. The affordable housing component has gradually been translated into rental housing irrespective of rent level. A separate rent-setting system has been created to cover costs in new construction. The land is allocated to potential developers as ownership or leasehold. According to the current land allocation policy, half of the land should be allocated to rental projects and leaseholds may be given to facilitate rental housing construction. Land allocation processes shall also stimulate competition to reduce housing costs. Land is sold or leased based on the market value of the land in its foreseen use. (Stockholm, 2015) Currently, the land allocation policy is under revision. The city has the vision to stimulate innovation and affordable housing business models and to use the municipal land allocation process to implement new ways of allocating housing units. (Stockholm, 2020) When implementing policy, the city has a multi-facetted role as a planning authority, landowner and property developer through its municipal housing companies. The municipality needs to balance these three roles and also consider city fiscal policy (including the norm that each project budget should be self-supporting).

Non-public actors are expected to produce the larger share of new housing. Non-public actors can be divided into commercial project developers, including those offering rent-to-buy schemes, shared ownership and cooperative rental, and non-profit actors such as foundations, cooperatives, and collaborative housing initiators. However, the Swedish non-profit housing sector is negligible in size. Further, cooperative rental and rent-to-buy/supported purchase schemes are being implemented only on a small scale (for example Bergsten & Granath Hansson, 2023). In NDS, this sector is limited to one building belonging to the Stockholm Cooperative Housing Company (SKB) that does not provide affordable housing in new construction. Instead, commercial project developers dominate. Inclusionary housing policies or similar vehicles designed to incentivise commercial project developers to build affordable housing are presently not applied in NDS, but there is a pilot project in another part of the city (Granath Hansson, 2020).

The city of Stockholm owns municipal housing companies that manage approximately 15% of the housing stock. The current norm in Sweden is 'good housing for all' and there is no social housing system or any other form of housing to which lower income groups have priority. According to a change of law on municipal housing companies in 2011, these companies shall act in 'a business-like manner', and not compete with private housing companies on unequal terms (Elsinga & Lind, 2013). The law was a consequence of negotiations between the Tenants' Union and the Federation of Property Owners related to EU regulation on fair competition, where the Tenant's Union were strongly advocating the 'good housing for all' position and refused to introduce the concept of means-tested social housing (in contrast to choices made in the Netherlands linked to the same EU debate). The meaning of 'business-like' has been interpreted in various ways by different companies (Grander, 2018). Although municipal housing companies do not have a formal role as social housing providers, they are in many locations taking on a larger social responsibility than other property owners (for example Borg, 2018; Grander, 2018). In NDS, municipal housing companies have built several projects and tried to keep rents at a moderate level although this has been a challenge considering the character of and goals for the area. Achieved rents are too high to be affordable to the larger part of the population. The serial housing concept 'Stockholmshusen' (Stockholm Houses) that is pointed out as the major affordable housing tool of municipal housing companies' is not used in NDS as its architectural expression is not compatible with the design program of the area.

4 Policies and Institutions that Influence Outcome

Provision of affordable housing in Sweden can be understood through the interaction between land use, housing and fiscal policy.

4.1 Land Use and Fiscal Policy

As described above, the City of Stockholm owns the major share of land within its jurisdiction and allocates parts of this land for housing. The municipal land allocation procedure is based on a formal land allocation policy but is in reality a multifaceted and complex case-by-case negotiation process between the developer and the municipality (Candel & Paulsson, 2023). Land allocation has been defined as an exclusive but time-limited right of a developer to negotiate a land lease or a land sale with the municipality (Caesar, 2016). When a land allocation has been made, negotiations of the land sales or lease agreement, the development plan (if such a plan does not already exist) and the implementation agreement take place. The implementation agreement regulates project specifics, such as project design and cost division between the developer and the municipality related to technical infrastructure. When the municipality and the developer agree on the terms of the development and the contracts are signed, the land is transferred to the developer. However, in case the municipality and the developer do not find a compromise between their interests at the end of the time limit in the land allocation contract, the municipality can either prolong or terminate the contract. The developer may also choose to withdraw from the project. In such cases, the land is usually offered to competing developers. The process has been criticised for being development rather than planning led (Zakhour & Metzger, 2018).

Land leases are priced based on the planned use of the land. As ownership projects are in general more profitable than rental, this means that if there is a choice between selling the land for ownership projects and leasing the land for rental units it would be easier to argue for ownership projects, as the choice of rental will result in losses to the municipality. However, in Stockholm lease prices are not differentiated between different types of rental housing, which entails that as long as the plot was planned to be leased to a rental housing developer, the municipality will suffer no extra loss should it decide to lease it to projects that wholly or partly contain lower rent apartments. The developer would however be less incentivised to participate as lower rent levels would not entail an extra subsidised land lease.

A number of factors speak for high land values in the area, for example the location and attractivity resulting in a potential for high prices and higher than usual rents in both commercial and residential space, as well as high environmental and architectonical demands. However, the city develops land also in less attractive areas, which might entail lower land values in the area than would otherwise be the case (cf. Graeme Guthrie, 2020). The principle that city development projects should be self-supporting and not burden city budgets has been applied throughout the implementation of NDS. However, in the latest budget the possibility to cross-subsidise projects such that only the total city budget, not each project budget, is balanced will be investigated with the goal of making projects in less attractive areas viable (Stockholm, 2023). This implies that development surpluses in higher-end areas might not be used to finance investment in those areas as before, but instead be used to finance for example public space or housing in other areas. Hence, there is no goal to introduce affordable housing schemes in the higher-end areas.

4.2 Housing Policy

As mentioned above, the city interprets affordable housing mainly as rental housing in general as there is no dedicated social housing sector. In the last decade, municipal housing companies have not had an explicit but rather an implicit role to play in rental housing provision; however, according to the latest city budget they are pointed out as important vehicles (Stockholm, 2023). In order to understand the environment in which such rental housing development takes place the rent-setting and apartment allocation systems must be understood.

Rents are collectively negotiated between property owners and the Tenants' Union mainly according to a separate rent-setting system for the first 15 years. This system was introduced to encourage new construction and allows for higher but stable rents. (Baheru, 2020). Some private property owners also chose to set their own rents without negotiating with the Tenants' Union, although they then risk a trial in the Rent Tribunal after which rents might be reduced to reflect rents in surrounding buildings of equal standard. In NDS, rents are set to reflect the high costs of project implementation, including land prices, contributions to infrastructure, greenery, innovative environmental standards etc., and in some projects more luxurious design. The result is that lower- and moderate-income households cannot afford to live in the area. Among Swedish municipalities there is a general uncertainty when it comes to effective measures to set and keep rents at lower levels over time in line with the Swedish rent-setting system (Granath Hansson, 2019). This institutional uncertainty might affect decisions to implement affordable housing schemes in development areas.

Further, allocation of apartments is made according to time-in-queue. Some groups at times get priority in queues, but often policy is not implemented in high-cost areas as the target households rarely can afford to pay the rents. However, social services sometimes rent apartments also in high-cost areas and then sublet to clients, typically when supply is scarce and there is no other solution. The goal to create social mix is therefore hard to attain from the outset as mainly higher income earners, and a few tenants assisted by the social services, are expected to rent the apartments when they are new. Instead, the area is expected to become increasingly affordable with time as the housing stock depreciates and rents are renegotiated. The end of the rent lock-in period of 15 years in parts of the housing stock then becomes a crucial point in time.

4.3   The Interaction Between Land Use, Fiscal and Housing Policies

So-called inclusionary housing policies that utilise the land use system to include affordable housing in otherwise market-rate projects with the goal of creating socially mixed housing areas are common in Europe (Granath Hansson et al., 2024). In these policies, the allocation of apartments to low- and medium-income households is central. De Kam et al. (2014, 397) point out: "Another important condition for the acceptance and societal support for IH [inclusionary housing] is usually the capacity of the housing system to retain the benefits of IH for eligible households exclusively, for a reasonable number of years". In contrast, Swedish policy tries to avoid reserving housing for special groups and explicitly shuns means-testing as part of the Swedish unitary housing system where equal treatment is a corner stone. There is also uncertainty regarding how to regulate rents for a predefined period of time as the Swedish rent-setting system is based on negotiations between property owners and the Tenants' Union and does not foresee such parallel rent-setting systems influenced by municipalities. Attempts to regulate rents through land lease agreements have been met with doubt (Granath Hansson, 2021). As a result, there is no model with the effect that lower income groups get priority in allocation of lower rent apartments that result from land policy models. There are however discussions and a few pilot projects where alternative allocation methods are tested, for example based on need rather than income (Granath Hansson, 2020).

Further, the municipality's three different roles as authority, landowner and developer create distinct and sometimes competing interests. As mentioned above, the municipality has the political goal to enhance affordable housing and social mix and has ordered its planning and development bodies to implement policies to such effect. However, parallel policies on design and environmental sustainability that drive development cost compete for priority and are much emphasised in NDS. Further, the municipality has to balance its project portfolio such that income exceed cost, and it can pay for its property development obligations related to planning, land readjustment, land sales and leases, as well as technical and social infrastructure. Here, losses incurred by decisions to lease land for rental housing development instead of selling land to the highest bidder might be seen as a form of subsidy. Although the municipality is aware of that few moderate- and mid-income households can afford initial rents, it is said that in time, as the housing depreciates and the rent-setting system has had its effects over a longer period of time, these groups will also gain access. However, as rent-setting is now made differently in different projects with variations between public and private property owners, this assumption might not hold in the future. Law requirements for municipalities to house vulnerable households do however at times work to include low-income households through the social services, usually in the municipal housing stock. Although rents paid by the municipality to house vulnerable groups are high compared to rents in the older stock, the municipality tends to save money on such arrangements compared to when shelter has to be negotiated in hotels and similar solutions.

The role of municipal housing companies in developing affordable housing is also limited. This partially has a background in legislation prohibiting municipal housing companies to compete with private property owners on unequal terms. One of the effects of the legislation is that municipalities cannot subsidze housing development and cross-subsidy between projects is also met with hesitation. There are attempts to create lower rent apartments, but in higher-end areas this has only limited results as land is valuable and municipal demands on design and sustainability drive costs. Moreover, the lower priced concept housing Stockholm Houses is not profitable to build, or accepted by the municipality as planning agency and landowner, in areas like NDS, but is built in suburbs where their architectural expression fits in.

5 Reflection

The choice not to include affordable housing in NDS might seem paradoxical in view of the city's vision to provide 'good housing for all' in socially mixed neighbourhoods. However, when combining the contradictory goals and incentives of the actors and institutions involved, the choice might be perceived as understandable or even rational.

In NDS and many other development areas throughout Sweden, housing and land policy are based on laws, rules and norms which contradict policy goals. When it comes to housing policy, four issues are decisive: First, it is uncertain how rents can be regulated for a certain period of time in line with the Swedish rent setting system. Second, an efficient allocation of the resulting affordable housing might not cope well with the Swedish unitary housing system in which means-testing is not implemented. Third, municipal housing companies have only limited incentives to build affordable housing and it is a challenge to create profitable or even zero-sum projects in high-end areas without subsidy. Fourth, serial affordable housing concepts like 'Stockholm Houses' are not deemed appropriate in areas with more elaborate design programs. Based on these four issues related to laws, rules and norms, it is not only difficult to build affordable housing from a financial and legal perspective, but also difficult to justify as there is no guarantee that the resultant housing will be inhabited by mid- or low-income households. As a result, incentives to build affordable housing are low already on the housing policy level. When combined with land policy, there are even more arguments speaking against it, as will be outlined below.

Today, granting municipal leasehold land to rental housing projects is the main affordable housing tool in land policy. However, leasing land leads to substantial losses of municipal income compared to selling the land to the highest bidder which has effects on the city's economy and possibilities to cover its costs for developing the area to the wished for standard (and perhaps in the future cross-subsidising projects in less attractive areas). As seen above, this subsidy does, however, not decrease rents to a level affordable to the larger part of the population. Both the loss of income and the relatively high rent levels limit the number of projects that are granted leasehold land, as the overall development area budget has to be balanced, and the effect of the subsidy is limited. As a result, models to incentivise private (and public) property developers to participate in inclusionary housing schemes have been used hesitantly. All in all, affordable housing land models might be seen as a complicated way to increase affordable housing supply, with small effects on the housing shortage as apartment allocation is not effective.

When the major goal is large quantities of newly built apartments and the land development process is rather complicated as it is, the hesitance to engage in more elaborate and time-consuming processes that would require new solutions, or even changes to the institutional set-up, is rational. There is also a clear balancing of priorities between the three dimensions of sustainability, where social sustainability might have to take a back seat compared to environmental and economic sustainability as these two sustainability dimensions are better defined and effects easier to identify in relation to the NDS development goals. The city of Stockholm has unique possibilities to realise affordable housing models through their planning monopoly and extensive land ownership, but it is clear that the wish to stem affordable housing shortage through land allocations has not been strong enough based on political and institutional reasons. It is probable that increased attention to affordable housing shortages and social mix policies is needed to evoke a change in actors' priorities and initialise institutional reform that takes care of at least some of the hurdles described above. In the 2024 city budget, politicians have included an assignment to city departments to develop new guidelines for housing provision, complementing the city master plan, and having a distinct impact on land policy. The guidelines shall be accompanied by a list of actions to be taken. It remains to be seen what changes might be introduced into Stockholm's land policy.

Although the Stockholm housing market is an extreme case in many respects, the hurdles highlighted above are also faced by other Swedish cities in their housing development efforts as not only the same laws, but also the same rules and norms, apply in most Swedish municipalities. Future research could investigate potential policy change related to the hurdles outlined above in various municipalities. An exchange of experiences between municipalities could prove useful for policy development.

References

Baheru, H. (2020). Hyressättning: *Prisets reglering vid bostadshyra.* Doctoral thesis. Stockholm University.

Bergsten, L., & Granath Hansson, A. (2023). Intermediary housing tenures in Sweden: Developers' response to inaccessible housing markets and its implications for tenant-buyers. *Nordic Journal of Urban Studies, 3*(1), 4–22.

Borg, I. (2018). Universalism lost? The magnitude and spatial pattern of residualisation in the public housing sector in Sweden 1993–2012. *Journal of Housing and the Built Environment, 34*(2), 405–424.

Caesar, C. (2016). *Municipal landownership and housing in Sweden: Exploring links, supply and possibilities.* Doctoral thesis. KTH Royal Institute of Technology.

Caesar, C., & Kopsch, F. (2018). Municipal land allocations: A key for understanding tenure and social mix patterns in Stockholm. *European Planning Studies, 26*(8), 1663–1681.

Candel, M., & Paulsson, J. (2023). Enhancing public value with co-creation in public land development: The role of municipalities. *Land Use Policy, 132,* 106764.

De Kam, G., Needham, B. & Buitelaar, E. (2014) The embeddedness of inclusionary housing in planning and housing systems: Insights from an international comparison, Journal of Housing and the Built Environment, 29, pp. 389–402.

Elsinga, M., & Lind, H. (2013). The effect of EU-legislation on rental systems in Sweden and the Netherlands. *Housing Studies, 28*(7), 960–970.

Guthrie, G. (2020). Incentivizing residential land development. *Housing Studies, 35*(5), 820–838.

Grander, M. (2018). *For the benefit of everyone? Explaining the significance of Swedish Public Housing for Urban Housing Inequality.* Doctoral thesis. Malmöuniversity.

Granath Hansson, A. (2019). Inclusionary housing policies in Germany and Sweden: The importance of norms and institutions. *Nordic Journal of Surveying and Real Estate Research, 14*(1), 7–28.

Granath Hansson, A. (2020). Land models for affordable housing: Drivers in the housing and planning systems in Sweden. In E. Hepperle, J. Paulsson, V. Maliene, R. Mansberger, A. Auzins, & J. Valciukiene (Eds.), *Methods and concepts of land management: Diversity, Changes and New Approaches* (pp. 199–209). vdf Hochschulverlag

Granath Hansson, A. (2021). Krav på bostäder med lägre hyra och sociala kontrakt i nyproduktionen. KTH Royal Institute of Technology, report TRITA-ABE-RPT-2135. Retrieved July 26, 2024, from https://kth.diva-portal.org/smash/get/diva2:1618711/FULLTEXT01.pdf

Granath Hansson, A., Sørensen, J., Tophøj Sørensen, M., & Nordahl, B. I. (2024). Contrasting Inclusionary Housing Initiatives in Denmark. How the past shapes the present. *Housing Studies.* Advance online publication. https://doi.org/10.1080/02673037.2024.2323607

Kalbro, T., & Lindgren, E. (2021). *Markexploatering.* Norstedts förlag

Sheiban, H. (2002). *Den ekonomiska staden: Stadsplanering i Stockholm under senare hälften av 1800-talet.* Arkiv förlag

Sidenbladh, G. (1981). *Planering för Stockholm 1923–1958.* Liber

Smas, L., & Schmitt, P. (2021). Positioning regional planning across Europe. *Regional Studies, 55*(5), 778–790.

Stockholm. (2015). Markanvisningspolicy. Retrieved July 26, 2024, from https://foretagsservice. stockholm/globalassets/delad-media/markanvisningspolicy-2015.pdf

Stockholm. (2018). Översiktsplan för Stockholms stad. Retrieved July 26, 2024, from https://vaxer. stockholm/siteassets/stockholm-vaxer/tema/oversiktsplan-for-stockholm/oversiktsplan-for-sto ckholms-stad-godkannandehandling-2020-10-03.pdf

Stockholm. (2020). Riktlinjer för bostadsförsörjning 2021–2024. Retrieved July 26, 2024, from https://edokmeetings.stockholm.se/welcome-sv/namnder-styrelser/kommunfullmaktige/mote-2020-11-02/agenda/bilaga-4-riktlinjer-for-bostadsforsorjning-2021-2024-slutlig-version-med-accepterade-andringarpdf?downloadMode=open

Stockholm. (2021). Program för hållbar stadsutveckling: Norra Djurgårdsstaden visar vägen mot en hållbar framtid. Retrieved July 26, 2024, from https://www.norradjurgardsstaden2030.se/doc/ sv/program-stadsutveckling-nds-2021.pdf

Stockholm. (2023). Budget 2024. Retrieved July 26, 2024, from https://start.stockholm/globalass ets/start/om-stockholms-stad/sa-anvands-dina-skattepengar/stadens-budget-ar-fran-ar/stockh olms-stads-budget-2024.pdf

Zakhour, S., & Metzger, J. (2018). From a "Planning-Led Regime" to a "Development-Led Regime" (and Back Again?): The Role of Municipal Planning in the Urban Governance of Stockholm. *The Planning Review, 54*(4), 46–58.

Anna Granath Hansson is a Senior Research Fellow at the Nordic research institute Nordregio and lecturer at KTH Royal Institute of Technology in Stockholm, Sweden. She holds a Ph.D. in real estate science based on her dissertation *Institutional prerequisites for affordable housing development: A comparative study of Germany and Sweden.* Her main research focus is housing, property development, land use policy and planning and she often takes a comparative perspective between cities and countries in Northern Europe. Before joining academia, Anna pursued a career in the real estate and construction industries in Europe.

Land Policy in Switzerland: A Renewed Interest in Public Landownership to Support Land-Use Planning

Jessica Verheij, Andreas Hengstermann, and Jean-David Gerber

Abstract Although land policy has a long history in Switzerland, the Swiss planning system does not provide a uniform understanding of what land policy entails. We propose that land policy should be viewed as a strategy to address the limitations of spatial planning. This chapter illustrates the strategic application of land policy and its potential to achieve policy goals through a case study of a large greenfield development in the city of Bern. Through leveraging public landownership and strategically combining planning instruments such as the municipal zoning plan and long-term ground leases, the city of Bern has successfully facilitated large-scale housing development. The municipality's active land policy strategy was crucial in achieving this outcome, demonstrating effective use of the planning instruments available. Moreover, this case illustrates the role of key actors in land policy in Switzerland. It demonstrates a renewed interest among Swiss cities in using public landownership to tackle significant special challenges.

1 Land Policy in Switzerland

Historically, Swiss federal land-use planning emerged as a public effort to protect agricultural land from the sprawl of expanding of cities. Through the use of strong zoning, planners were tasked with defending the countryside and its agricultural

J. Verheij (✉)
École Polytechnique Fédérale de Lausanne, Lausanne, Switzerland
e-mail: jessica.verheij@epfl.ch

A. Hengstermann
Norwegian University of Life Sciences, Ås, Norway
e-mail: andreas.hengstermann@nmbu.no

J.-D. Gerber
University of Bern, Bern, Switzerland
e-mail: jean-david.gerber@unibe.ch

© The Author(s) 2025
T. Hartmann et al. (eds.), *Land Policies in Europe*,
https://doi.org/10.1007/978-3-031-83725-8_12

landscapes, which hold a strong symbolic dimension within Swiss national identity. This historical precedent provides insight into the specifics of land-use planning in Switzerland and, indirectly, the rise of land policy. First, the central objective of separating buildable from non-buildable land created a narrow definition of the scope of land-use planning. This led to, second, an abrupt separation between land-use planning and other public policies, such as housing and transportation, which are dealt with in different legislations under the responsibility of different federal offices. Third, the comparatively weak position of the Federal Spatial Planning Office, which only marginally deals with the urban environment, left the main responsibilities in the hands of the municipalities. Fourthly, a very strong definition of property rights implying limited expropriation possibilities, coupled with, fifthly, semi-direct democratic principles that require a popular vote for each change in land-use plans, greatly limited the flexibility of Swiss land-use planning. These characteristics explain the difficulty of present-day Swiss land-use planning to tackle present-day challenges such as urban densification or the preservation of affordable housing, which necessitate strong public intervention.

In Switzerland, land policy is to be understood as a strategy to overcome these limitations. Through land policy, public authorities strengthen their position in implementing spatial planning objectives. Land policy has a long history in Switzerland, as, prior to 1970s, most larger municipalities used to acquire land to influence urban development. However, with the rise of neoliberalism in the late 1980s, many municipalities sold their land to generate instant cash (e.g. to pay back debts). During this period, several federal service providers, such as the Swiss Federal Railways and the Swiss Post, were privatised, demonstrating a transfer of public landownership to semi-public entities. Yet since the 2000s, the general expectation that public authorities should become more output-oriented and efficient (in line with New Public Management), has prompted a renewed interest in land policy (Gerber, 2016). This has led to a greater focus on public ownership to accompany spatial planning at the municipal level and the development of specific strategies for different levels of government (acquisition funds, ban on the sale of public land, long-term ground leases, pre-emption rights, public–private partnerships, etc.) (Debrunner & Hartmann, 2020).

All in all, land policy in Switzerland has a very pragmatic connotation. As it falls under the responsibility of each municipality, there is no uniform understanding of what it means. At the national level, the term land policy (in German *Bodenpolitik*; in French *politique foncière;* and in Italian *politica del suolo*) does not appear explicitly in any federal law or official government document. Land policy[1] is, therefore, not a (public) policy, but a strategy. We define land policy as the set of public and private law instruments that municipalities combine to reinforce the implementation of spatial planning objectives, which directly reflect the will of the Swiss voting population.

---

[1] Another understanding of land policy relates to the protection of soil quality. This policy, which is in its early stages of development, is promoted through research funding programs and lobbying association in the natural sciences. We will not deal with this definition of land policy in this chapter.

As the implementation of these objectives often proves difficult, especially in the already built environment, municipalities make use of land policy strategies.

To illustrate how land policy in Switzerland functions in practice, we present the case study of *Viererfeld*. Located in Bern, it is one of the city's last large-scale, centrally located areas yet to be developed (see Fig. 1). The case study demonstrates how the municipality of Bern devised an active land policy strategy using policy instruments deriving from the institutional regime in place.

Fig. 1 Viererfeld in Bern, Switzerland (map: Kleiner, Jehling 2024, Aerial Imagery Provider © Esri, swisstopo, Planex, Maxar, Microsoft)

2 Understanding Land Policy in Switzerland: The Case of *Viererfeld* in Bern

The development of *Viererfeld/Mittelfeld* (henceforth *Viererfeld*) is located on a greenfield area in the city of Bern. The fifth largest city in the country, with a population of 144,000 (as of 2022), Bern is the capital of Switzerland as well as the capital of the Canton of Bern. In addition, Bern is the centre of several regional organisational levels (including the agglomeration of Bern), though these play a marginal role in planning and land policy.

Bern's UNESCO World Heritage-protected cityscape is characterised by medieval arcades, most of which were built after the great city fire in 1405. Since then, through the development of new residential neighbourhoods, the city has been steadily growing outwards. Since the early 2000's, the city's population has been growing, due in part to the widening availability of jobs in the city and its neighbouring municipalities (Stadt Bern, 2022). However, these growth processes and the resulting increase in commuter movements present significant challenges to the city's current spatial development.

2.1 Understanding Bern's Housing Market to Understand the Importance of Viererfeld

The city's growing population is considered a main challenge in many spatial domains, including housing. Switzerland is a 'nation of renters' where over 60% of households live in rental housing (BfS, 2022). The city of Bern is no exception, with only 17% of the population owning their own home. Yet over the last few years, rental prices have risen disproportionately compared to overall consumer prices (Stadt Bern, 2018b). Bern currently experiences a vacancy rate of 0,45% (approx. 350 housing units) (Stadt Bern, 2018a), indicating an excess demand in relation to supply. Since 2003, rental prices for housing in the city of Bern have increased by 24,3%, similar to overall trends on the national level (Stadt Bern, 2018b). The rent burden is, however, distributed rather unequally. Low-income classes in Switzerland spend over 35% of disposable income on rental costs, against a national average of just 20%. Moreover, this share has been rising for low-income classes, while the share for high-income households has remained stable over the last 20 years (BWO, 2022). A further challenge relates to the inefficient allocation of existing housing. With less than 25% of all rental units containing more than three rooms, a large share are occupied by small households of only one or two individuals. Families are, therefore, experiencing increased difficulties in finding appropriate housing within the city (Stadt Bern, 2018b).

With a housing stock of circa 7000 units, Bern's housing market is dominated by private for-profit actors and individual private actors who own over 75% of housing units. The remaining units are owned by non-profit actors and public actors, the latter owning only 2% of the housing stock. In recent years, ownership structures have shifted mostly towards institutional actors. Newly-built housing developed between 2004 and 2015 is owned primarily by developers (both for-profit and non-profit) and other legal entities, with only a marginal number of housing being developed by private individuals or public actors (Stadt Bern, 2018a).

Bern's housing strategy is mainly directed towards increasing the housing supply within the existing urban fabric (via inward development or densification), focusing specifically on family-appropriate housing and an increase in affordable or non-profit housing. Until 2030, 50% of all newly developed housing units are to be either affordable or non-profit (Stadt Bern, 2018b). Furthermore, the city's spatial development plan stresses the need to increase the housing supply: it sets a target of 8'500 additional housing units (+11%) by 2030 and identifies potential densification areas across the city (Stadt Bern, 2016). The housing strategy and the spatial development plan both highlight densification and inner-city development. These strategies mirror the political orientation of the city's left-green government, by focusing on providing affordable housing and recognising the need for state intervention. Nevertheless, the city has limited leeway for public intervention in a housing market dominated by for-profit actors. Therefore, its housing strategy recognises the potential for acting strategically by combining several available instruments, such as planning regulations (including but not limited to zoning), long-term ground leases, and social policy.

Here, the municipal Fund for Land and Housing Policy (*Fonds für Boden- und Wohnbaupolitik*) plays a crucial role. It was created in 1985 after the electorate voted in favour of more active land policy and intervention in the housing market. This fund, integrated into the municipal real-estate department, serves as a primary instrument for implementing housing policy through three key actions (ISB, 2021). First, it actively purchases real estate within the city to increase its portfolio. Second, it allocates existing city-owned housing stock to low-income and other marginalised groups. Third, it issues ground leases to both for-profit and non-profit housing developers to influence housing development.

The *Viererfeld* case study examined in this chapter is central to Bern's housing and spatial strategy. Not only does it represent one of the last undeveloped plots in the city, but due to its ownership structures, it also provides the opportunity to closely control its development.

2.2   The Viererfeld Project

Viererfeld is a greenfield area located in the northern part of the city. Despite its proximity to an existing residential neighbourhood, the area feels secluded due to its location on an elevated plateau between two major roads. Over the last few decades, the land has been used mainly for agriculture. Today, *Viererfeld* also provides some public uses, including allotment gardens, a sports field, a mountain bike track and, more recently, a temporary refugee centre. The area is accessible by public transport via direct buses to Bern's main train station and is furthermore well connected to the recreational forest on the northern side of the city.

In the 1960s, the canton of Bern acquired the land from Bern's civic corporation, while seeking to develop a new university campus on the fringe of the city. Yet as these plans fell through, the plot presented new possibilities. Now owned by the cantonal government, its potential for housing development was recognised, and *Viererfeld* became a recurring topic in the city's political debates on urban development.

At the beginning of the 2000s, the city of Bern was experiencing an increasing lack of available housing within its city boundaries. This led to growing calls for developing *Viererfeld* into a new residential neighbourhood. Nevertheless, in 2004, the first attempt to zone *Viererfeld* as buildable land was narrowly rejected by the municipal electorate (Table 1). In most Swiss municipalities, including Bern, the electorate can approve or reject changes to the zoning plan, making public acceptance a major factor in urban development. Over the years, housing became an increasingly pressing issue across all market segments. Consequently, the planning process was revived in the 2010s, with a new zoning plan finally being approved in 2016. Along with the zoning plan, the electorate also approved the city's purchase of half of the *Viererfeld* plot from the canton. Since then, *Viererfeld* has been jointly owned by both the canton and city of Bern.

Viererfeld's proposed zoning plan complements the city's regular zoning plan, in line with so-called special land-use plans. Derived from cantonal planning law, these plans enable detailed regulations of new development through specific land-use and building rules. The *Viererfeld* plan, approved in 2016, changes land use from agricultural to housing and mixed-use, and generally divides the land into two parts: the city-owned half is to be developed into a high-density compact development of 1,140 housing units. The canton-owned half is to be developed into public green space. It furthermore stipulates that a minimum of 50% of the housing on *Viererfeld* is to be developed as non-profit. This regulation builds on and extends Art. 16b of the Building Code of the city of Bern (*Bauordnung der Stadt Bern*), which mandates that at least one third of all housing in newly zoned areas be developed as affordable, based

Table 1 Overview of public votes on the development of Viererfeld

Year	Legal document	Core content	Result
2004	Zoning plan	Change of land use from agriculture to housing	Rejected (48.29%)
2016	Zoning plan	New zoning plan Budget to purchase plot Budget to continue planning	Accepted (53.02%)
2023	Credit for plot development	Project plan, including public investment credit for servicing the land (streets, greenspace, playgrounds)	Accepted (64.08%)
2023	Ground lease contracts with developers	Ground-lease to non-profit developer	Accepted (75.77%)
		Ground-lease to for-profit developer	Accepted (66.93%)

on cost-rent principles. Recognizing its powerful position to control the development of *Viererfeld* through its property rights, the city of Bern saw the opportunity to increase the share of non-profit housing to 50% (Stadt Bern, 2020a).

After the rejection of the first zoning plan in 2004, the city understood there was strong political resistance against the development of *Viererfeld*. This resistance mainly originated from a desire to preserve the fertile agricultural land of *Viererfeld*, which also serves as one of the few open green spaces with a view of the Alps in this part of the city. To ensure public approval, the city incorporated extensive green space development into its new plans. The approved zoning plan reserves half of the land for green space, featuring a publicly accessible park with playgrounds and sport facilities, allotment gardens, and several ecological compensation areas (Stadt Bern, 2020b). Hence, the *Viererfeld* project not only adds a significant amount of new housing, but also creates a large new urban park.

Property rights constitute a fundamental factor in *Viererfeld*'s planning process. As both the canton and the city own the land, they jointly influence its development through their property rights. For the green space development, the canton-owned half has been given out in ground-lease to the city for 50 years. After this period, the canton will have the possibility to reconsider its land use. Similarly, the city-owned half is given out to housing developers, of which 50% is to be non-profit and provide housing based on cost-rent principals. In this model, rents are calculated based on construction and maintenance costs, including capital costs and a 2% return on investment, without maximizing profits. Using ground lease contracts, with an average duration of 80 years, the city maintains control over the type of housing to be developed in *Viererfeld*, while safeguarding the possibility of reconsidering land uses in the future. Moreover, leasing buildable land prevents speculation, as the ground lease contract grants the city a pre-emption right to purchase buildings from developers at a predefined price (see also Balmer & Bernet, 2015). Thus, through landownership and ground lease contracts, public authorities ensure control over *Viererfeld*'s development both now and in the future.

2.3   Land Policy and Swiss Semi-Direct Democracy

In Switzerland, zoning plans and building codes determine the type and extent of permissible land use. As such, they are binding to both public authorities and private landowners. Changing these regulations typically requires approval by the electorate through a popular vote (Kanton Bern, 2018a, 2018b, 2018c). Given Switzerland's political culture of involving as many stakeholders as possible to maximise acceptance, popular votes are usually preceded by intensive political debates aiming for compromises between all parties. As a result, while Swiss politics can be rather slow, the decisions are generally very robust (Schmid et al., 2021; Solly, 2021; Wicki et al., 2022). The approval rates of various proposals for the development of *Viererfeld* (Table 1) indicate the controversial nature of this project. While similar projects in the city of Bern typically enjoy approval rates of over 70%, the zoning of *Viererfeld* faced initial resistance given its former agricultural use and recreational character. The city secured sufficient support from the electorate only after prioritizing green space development and a high share of non-profit housing.

3 Actors

The *Viererfeld* project involves key actors of Bern's spatial development and exemplifies typical actor constellations in Swiss planning processes. Bern's civic corporation owned the land until 1964, when it was sold to the cantonal government. As of 2016, the municipality of Bern owns half of the land, while the canton retains ownership of the other half. Each of these actors pursue their own land-related strategies.

3.1 Civic Corporation

Civic corporations (*Burgergemeinden*) are a particular but important landowner in Switzerland. These are public-law institutions that have their roots in the Middle Ages and still exist in parallel to the political municipalities. In Bern, the civic corporation is among the largest landowners, owning almost 30% of the land which includes a diverse portfolio of real estate assets, ranging from forests to cultural venues, and from housing projects to heritage buildings. Bern's civic corporation has been pursuing a strategy of not selling its land for over a century. Instead, the land is released directly for charitable purposes (through long-term ground leases) or managed in an active portfolio (Gutjahr, 2018). Much of the financial profit is donated to public purposes such as libraries and cultural institutions. The sale of *Viererfeld* to the canton was an exception to its general strategy of not selling land. Generally, it has a legal obligation to manage its land portfolio based on long-term objectives, quite independently from shifting political priorities and concerns (Gerber et al., 2011).

3.2 Cantonal Government

The cantonal government of Bern pursues a land banking strategy to provide space for economic development and large infrastructure projects. However, it is not directly involved in housing development, as this falls under the responsibility of the municipalities. Land is acquired or made available for companies that require large, contiguous areas, or to promote local economic development via designated "development hubs" (*Entwicklungsschwerpunkte*). Land may also be acquired when needed for large infrastructure projects, such as roads or airports. This included the purchase of *Viererfeld*, which was initially intended to be developed into a new university campus in the 1960s.

3.3 Municipal Government

The municipal government of Bern plays the role of both landowner and regulator in the development of *Viererfeld*. From 1959 to 1984, the city of Bern pursued a direct and active land banking strategy. Under this approach, the local government was tasked with acquiring land for public services and housing projects. However, in 1985, this strategy shifted to allow for greater flexibility and quicker responses to market conditions. Since then, the city has established a municipal company called the Fund for Land and Housing Policy. While the company operates under the political supervision of the executive council, it manages its portfolio independently. Currently, the Fund manages about 3000 residential units and 450 commercial units, in addition to leasing nearly 400 plots of land (ISB, 2021).

3.4 Non-landowning Actors

The electorate has a key role in Swiss planning processes. Based on principles of semi-direct democracy, the electorate generally votes on all legislative matters, including land-use regulations such as the Spatial Planning Act, zoning plans, and special land-use plans. Thus, beyond public participation processes, the Swiss electorate has a final say in approving or rejecting new laws and plans. Consequently, actors involved in spatial development, such as planning authorities or private developers, must carefully consider the likelihood of a development plan to be approved by the voting majority, often resulting in lengthy decision-making and planning processes.

Swiss law narrowly defines who can legally appeal against a development project. To file an appeal, one must both live in the direct vicinity (usually not more than 50 m) and experience concrete material damage due to the development in question. In the case of the *Viererfeld* project, its secluded location on top of a plateau and surrounded by infrastructure meant that only a few actors were eligible to appeal. As a result, the potential to appeal did not significantly affect the project's progression.

4 Institutions of Land Policy in Switzerland

4.1 Land-Use Planning

The planning system in Switzerland reflects the country's federalist political structure, comprising three levels of government: the federal state, the cantons, and the municipalities. Although the federal state has the highest level of authority, its role is mainly restricted to creating frameworks and establishing guiding principles for spatial development, ensuring coherence and integration on a national level. This rather passive role at the federal level is a direct consequence of the political debates surrounding the development of the Federal Spatial Planning Act in the 1960s and 1970s. In 1969, spatial planning was established as a public competence under the Swiss Constitution (today: Art. 75 Swiss Constitution) after approval by the voting majority. The approval process, however, was marred by significant conflict between left- and right-wing parties (Knoepfel & Nahrath, 2007). Overall, it took the parliament ten years to agree on the Federal Spatial Planning Act, the formulation of which makes several important concessions to the federalist circles and property rights lobbies. The law's final version did not include any property-based instruments and reinforced the power of the cantons to define their own spatial planning policy at the cost of the federal state. Hence, the Federal Spatial Planning Act can be considered a framework law, which, together with federal conceptual and sectorial plans, sets out a policy framework while giving sufficient leeway to the cantons to set out their own planning regulations (Bühlmann et al., 2011).

Given the variation in cantonal planning laws, the possibilities and responsibilities of municipalities in spatial planning vary accordingly. Generally, municipal governments are responsible for local land-use plans, which require approval by cantonal authorities. In Bern, spatial planning relies on a general building code and a zoning plan. This zoning plan is the primary instrument for regulating spatial development and is the only spatial plan with legally binding consequences for landowners. It typically divides the municipal territory into buildable and non-buildable land, further subdividing buildable zones into various use zones. In addition to the general zoning plan, most cantons, including Bern, allow municipalities to define special land use plans (*Sondernutzungspläne*) for specific areas. These plans can either deviate from or complement the zoning plan and building code, providing flexibility to regulate development projects with detailed, context-specific rules. In Bern, these plans are the preferred instrument for regulating inner-city development (Kanton Bern, 2018a, 2018b, 2018c). Like general zoning plans, special land use plans must also be approved by the electorate.

4.2 Property Rights

Landownership in its current form has been legally defined and protected nationwide since 1912, when the Swiss Civil Code (CC) came into force. Since then, "the owner of an object is free to dispose of it as he or she sees fit within the limits of the law" (Art. 641 para. 1 CC). According to the legal definition, landownership is considered, as three-dimensional, encompassing not only the land but the airspace above and the soil below it, insofar as there is an interest in exercising these rights (Art. 667 CC). In addition, landownership includes all permanently attached objects, such as buildings and trees (Art. 667 para. 2), as well as their products and fruits (Art. 643).

The definition of property rights has remained very stable. However, two notable changes have been made since the adaption of the original Civil Code. *First,* in 1963, condominium ownership was incorporated into the CC (Art. 712a-712t CC), allowing for ownership of parts of a building with similar rights as those of landowners. *Second,* in 1969, property rights were explicitly protected in the Swiss Constitution, coinciding with the establishment of spatial planning as a federal competence. The constitutional amendment stated: "The right to own property is guaranteed. The compulsory purchase of property and any restriction on ownership that is equivalent to compulsory purchase shall be compensated in full." (Today: Art. 26 para. 1–2 Swiss Constitution). While this constitutional provision reaffirmed existing Civil Code protections, it was part of a strategic compromise: to ensure a majority for the approval of spatial planning as federal competence, it was deemed necessary to simultaneously elevate the protection of property rights to constitutional status (Bühlmann et al., 2011; Hengstermann & Gerber, 2015).

4.3   Land Register and Public-Law Restrictions

The geodetic definition of a plot of land, along with the associated private law rights, are documented in the land register (Art. 942 CC). Following the adoption of the Civil Code, the national legislator tasked the cantonal administration with surveying the entire territory and documenting all land parcels. The land register is publicly accessible (Art. 970 para. 2 CC), but certain information, such as easements and purchase history, can only be requested if there is a legitimate interest (art. 970 para. 1 CC). However, data on land and infrastructure values, including public valuations (e.g. based on a standard land value reference system like in Germany) or purchase prices, is not publicly available.

In addition to ownership data, the public-law restrictions on landownership are documented in the Cadastre of Public Law Restrictions on Landownership.[2] These restrictions can arise from various policies including those related to spatial planning, transportation, and nature conservation. This information is crucial for landowners, and since 2008, it has been standardised and made centrally available.

For the *Viererfeld* case, the land register extract identifies three separate parcels (1192, 2341, and 2750), divided between municipal and cantonal ownership. The municipal parcel is owned by the municipal Fund for Land and Housing policy. The cantonal parcel is owned by the Canton of Bern, or more precisely, the Office for Land and Buildings, which has granted a ground lease to the City of Bern. The excerpt also details the planning regulations in place for *Viererfeld*, as defined through the special land-use plan.

---

[2] In German: *Kataster der öffentlich-rechtlichen Eigentumsbeschränkungen (ÖREB); in French: restrictions de droit public à la propriété foncière (RDPPF); in Italian: restrizioni di diritto pubblico della proprietà (RDPP).*

4.4 Specific Property Institutions in Swiss Legislation

The Swiss property regime includes some notable and specific rules, including significant restrictions on land transactions in specific cases. For instance, since 1983, the acquisition of land by foreign nationals has been restricted.[3] Additionally, the Peasant Land Act of 1991 limits the purchase of agricultural land by organizations and non-farmers.[4] Land that has been continuously used for 10–30 years can be legally transferred from possession to ownership (Ar. 661-662 CC). Swiss law also regulates the formation of new land, such as through alluvial deposits or changes in the river courses (Art. 659-660b).

5 Reflection

Land-use planning in Switzerland has historically focused on separating buildable from non-buildable land to protect agricultural land from urban sprawl. However, spatial challenges grow as available space within cities becomes increasingly scarce, necessitating more strategic and proactive intervention. Municipal planning authorities increasingly shape spatial development by combining public and private law instruments. In Bern, the rising demand for new housing, combined with a commitment to compact city development, required proactive municipal action. The *Viererfeld* project exemplifies the revival of land policy in Switzerland.

While housing policy has not been traditionally seen as an integral part of spatial planning, these two domains often overlap in the everyday practice of municipal planners. Traditional zoning which focuses on separating buildable from non-buildable land primarily aims to restrict and restrain urban growth. It is rather passive and works well to defend the status quo. However, the city of Bern, like most large Swiss cities, recognizes the need to go beyond mere zoning to achieve its policy objectives. With the *Viererfeld* project, the city took a proactive approach to creating space for housing, leveraging the power of property rights to do so.

The *Viererfeld* case is emblematic of the power of planning authorities when combining zoning with property rights. Through public landownership and long-term ground leases, the city of Bern ensured a particularly high share of non-profit housing (50%) as well as the development of green space. It further implemented ambitious energy and mobility regulations and fostered a mix of housing developers. In this case, Bern acts as both landowner and planning authority with the power to restrict and regulate land uses. This double role allows planning authorities to strengthen their position by mobilizing property rights. By strategically using private-law instruments, the city can mitigate the relatively weak influence of land-use planning and enforce spatial planning objectives.

---

[3] Bundesgesetz über den Erwerb von Grundstücken durch Personen im Ausland (BewG) vom 16. Dezember 1983. SR 211.412.41.

[4] Bundesgesetz über das bäuerliche Bodenrecht (BGBB) vom 4. Oktober 1991. SR 211.412.11.

When municipalities cannot rely on public ownership, they can supplement the limited scope of zoning plans by making use of special land-use plans, which allow for deviations from or additions to standard zoning regulations on a project-specific level. While these special land-use plans offer increased flexibility in planning processes, they remain reactive in nature. It is through the combination of property rights that planning authorities can proactively shape new development.

The *Viererfeld* case also illustrates the interplay between different levels of government in spatial development. While the municipality of Bern is responsible for the municipal zoning plan, the designation of *Viererfeld* as buildable land required approval by the cantonal government. At the same time, the electorate plays a crucial role, as any significant changes to the zoning plan must be approved by the public. Development only proceeded after the electorate approved the new zoning plan of *Viererfeld* in 2016. This interplay between different levels of government and the electorate exemplifies how spatial planning, like many other public policies in Switzerland, involves finding compromises and robust solutions that appeal to the majority.

Due to the federalist nature of the planning system and the lack of a unified definition of land policy, this case study may not be fully representative for the Swiss context. Nevertheless, it is typical of the way larger municipalities tackle present-day spatial development challenges. It clearly shows how a city like Bern strategically used municipal property rights to steer spatial development toward creating a new, compact, and green residential neighbourhood at the fringe of the city. The case study illustrates the renewed interest in land policy, especially in large cities seeking to address significant spatial challenges related to population growth, housing supply, and affordability.

References

Balmer, I., & Bernet, T. (2015). Housing as a common resource? Decommodification and self-organization in housing—examples from Germany and Switzerland. In M. Dellenbaugh, M. Kip, M. Bieniok, A. Müller, & M. Schwegmann (Eds.), Urban Commons: Moving Beyond State and Market. Birkhäuser. https://doi.org/10.1515/9783038214953-012

BfS. (2022). Bau- und Wohnungswesen 2020. Bundesamt für Statistik

Bühlmann, L., Haag, H., Jud, B., Kissling, S., & Spori, N. (2011). Einführung in die Raumplanung. VLP-ASPAN Schweizerische Vereinigung für Landesplanung

BWO. (2022). Durchschnittliche Mietbelastung nach Einkommensklassen 2006–2020. Bundesamt für Wohnungswesen. Retrieved September 6, 2024, from https://www.bwo.admin.ch/bwo/de/home/Wohnungsmarkt/zahlen-und-fakten/mietbelastung.html

Debrunner, G., & Hartmann, T. (2020). Strategic use of land policy instruments for affordable housing—Coping with social challenges under scarce land conditions in Swiss cities. *Land Use Policy, 99*(August), 104993. https://doi.org/10.1016/j.landusepol.2020.104993

Gerber, J.-D. (2016). The managerial turn and municipal land-use planning in Switzerland—evidence from practice. *Planning Theory and Practice, 17*(2), 192–209. https://doi.org/10.1080/14649357.2016.1161063

Gerber, J.-D., Nahrath, S., Csikos, P., & Knoepfel, P. (2011). The role of Swiss civic corporations in land-use planning. *Environment and Planning A, 43*(1), 185–204. https://doi.org/10.1068/a43293

Gutjahr, M. (2018). Die Überbauung des Schwabguts 1957–197—Die Burgergemeinde Bern als Akteurin in der städtischen Wohnraumpolitik. *Berner Zeitschrift Für Geschichte, 80*(3), 3–41.

Hengstermann, A., & Gerber, J.-D. (2015). Aktive Bodenpolitik—Eine Auseinandersetzung vor dem Hintergrund der Revision des eidgenössischen Raumplanungsgesetzes. *Flächenmanagement und Bodenordnung, 2015*(6), 241–250.

ISB. (2021). *Fonds für Boden- und Wohnbaupolitik—Gesamtstrategie und Teilstrategien.* Immobilien Stadt Bern.

Kanton Bern. (2018). *Eine neue Generation der Nutzungsplanung im Kanton Bern—Schlussbericht.* Kanton Bern.

Knoepfel, P., & Nahrath, S. (2007). Environmental and Spatial Development Policy. In *Handbook of Swiss politics* (pp. 705–733). Neue Zürcher Zeitung Publishing.

Schmid, F. B., Kienast, F., & Hersperger, A. M. (2021). The compliance of land-use planning with strategic spatial planning—Insights from Zurich Switzerland. *European Planning Studies, 29*(7), 1231–1250. https://doi.org/10.1080/09654313.2020.1840522

Solly, A. (2021). Land use challenges, sustainability and the spatial planning balancing act: Insights from Sweden and Switzerland. *European Planning Studies, 29*(4), 637–653. https://doi.org/10.1080/09654313.2020.1765992

Stadt Bern. (2016). *Stadtentwicklungskonzept Bern 2016.* Stadtplanungsamt der Stadt Bern.

Stadt Bern. (2018a). *Grundlagenbericht zur Wohnstrategie.* Stadtplanungsamt der Stadt Bern.

Stadt Bern. (2018b). *Wohnstrategie Bern—Wohnstadt der Vielfalt.* Stadtplanungsamt der Stadt Bern

Stadt Bern. (2020a). *Masterplan Viererfeld/ Mittelfeld.* Stadtplanungsamt der Stadt Bern

Stadt Bern. (2020b). *Wettbewerbsprogramm Arealentwicklung Viererfeld/Mittelfeld.* Stadtplanungsamt der Stadt Bern

Stadt Bern. (2022). *Statistik der Unternehmensstruktur 2020.* Statistik Stadt Bern

Wicki, M., Hofer, K., & Kaufmann, D. (2022). Planning instruments enhance the acceptance of urban densification. *Proceedings of the National Academy of Sciences, 119*(38), e2201780119. https://doi.org/10.1073/pnas.2201780119

Jessica Verheij is a researcher at École Polytechnique Fédérale de Lausanne (EPFL) in Switzerland. She holds a PhD in Geography from the University of Bern. Her research focuses on the planning for urban sustainability in European cities. She is interested in understanding the political processes behind the implementation of sustainability goals and the role of property institutions in urban development. She previously conducted research in Lisbon (Portugal) and holds master's degrees in Urban Planning (KTH Royal Institute of Technology) and Human Geography (University of Amsterdam).

Andreas Hengstermann is an Associate Professor with the Department of Urban and Regional Planning (BYREG) at the Norwegian University of Life Sciences (NMBU). He holds a Ph.D. in Geography from the Institute of Geography of the University of Bern (CH). He was educated as a planner (Dipl.-Ing. Raumplanung) at the TU Dortmund University (DE) and the Universidad de Huelva (ES). Furthermore, he did postgraduate studies in public law (DAS Law). His primary research interest lies in understanding the essential role of property rights in shaping spatial development. His research includes comparative studies across different geographical contexts, political systems, and legislations. He current serves as Vice-President International Academic Association on Planning, Law, and Property Rights.

Jean-David Gerber is an associate professor at the Institute of Geography of the University of Bern. He holds a postgraduate degree in Urban Development, Resources Management and Governance (2004) from the University of Lausanne and a PhD in Public Administration (2005) from the Swiss Graduate School of Public Administration (IDHEAP). His current research focuses on land policy, housing policy and social sustainability with a special emphasis on commons.

Land Policy in the Netherlands: An Ambiguous Utopia on the Move

Edwin Buitelaar, Martijn van den Hurk, Jasper Lebbing, Peter Pelzer, and Lilian van Karnenbeek

Abstract The Netherlands was traditionally lauded for its planning system, including its land policies. In this chapter, we argue that the picture has always been more nuanced and local planning practice has always been more pragmatic; it was an ambiguous utopia at best. Moreover, the Netherlands has experienced a transition over the last thirty years from active land policy with land ownership of municipalities towards facilitative (or passive) land policy in which developers typically own land and are active in initiating zoning changes. The latter can also be couched as 'institutional entrepreneurship' of landowners, where distinctions between the private and public become fuzzy, sometimes in problematic ways. We illustrate this conundrum through the case study of *Rijnenburg*, a polder close to the city of Utrecht, which has been a 'planning battle scene' for over thirty years, with different claims—housing, renewable energy, climate adaptation—and its concomitant representatives competing for prevalence. We conclude that the current system of public–private fuzziness is particularly vulnerable in times of a polycrisis. This predicament calls for a thorough consideration of long-term consequences, for instance in relation to climate adaption and urban development.

E. Buitelaar (✉) · M. van den Hurk · J. Lebbing · P. Pelzer · L. van Karnenbeek
Utrecht University, Utrecht, The Netherlands
e-mail: e.buitelaar@uu.nl

M. van den Hurk
e-mail: m.h.h.vandenhurk@uu.nl

P. Pelzer
e-mail: p.pelzer@tudelft.nl

L. van Karnenbeek
e-mail: l.j.vankarnenbeek@uu.nl

© The Author(s) 2025
T. Hartmann et al. (eds.), *Land Policies in Europe*,
https://doi.org/10.1007/978-3-031-83725-8_13

1   Understanding Land Policy in the Netherlands

We are facing a polycrisis: climate change, international migration, unaffordable and lack of housing, and unsustainable food production currently all come to the fore and interact with each other. Land is the stage on which the polycrisis needs to be dealt with. It is no surprise then that decision-making about land use is more contentious than ever. Consequently, land policy is more relevant than it has ever been.

This is also the case in the Netherlands, where land is scarce and many sectors compete for this land—housing, agriculture, nature, renewable energy etc. The land-use plan[1] is the legally binding document in which spatial decisions are formalised, and on the basis of which planning permissions are granted or rejected. Although it is a cornerstone in the planning process, this document follows and formalises decisions made by stakeholders, rather than that it guides them. The actual decisions follow a less formalised process infused with power play.

Private and public landowners are key stakeholders in the spatial decision-making processes. Getting hold of the land appears to be pivotal for the course of land-use planning and development, also for local authorities. Private developers on the one hand put pressure on the planning process to get land rezoned for housing. On the other hand, public actors also need to consider long-term interests, including climate adaptation. It is a balancing act along, at least, three dimensions. This chapter aims to use the case of *Rijnenburg* (Utrecht), a large and contentious greenfield site, to show these three dimensions, that is (1) how formal land-policy instruments are used and embedded in informal land development practices, (2) how long-term and short-term interests are traded off, and (3) how public law and private law are combined in the Netherlands (see Fig. 1).

The chapter is structured as follows. First, we describe the history and basic features of Dutch land policy (next section). After that we discuss three components through which we analyse *Rijnenburg* and Dutch land policy: the resource, the actors, and the institutions. We conclude with several reflections.

Land policy in the Netherlands can be described as a behavioural pattern, usually followed by municipalities, aimed at developing land for urban purposes. Compared to other countries, land policy in the Netherlands has a relatively narrow description. First, it is not synonymous to land-use policy or spatial policy, but *instrumental* to those two as it is focused on bringing about spatial and land-use policy goals through (urban) *change* (hence not the goals aimed at preserving the status quo). Moreover, as Needham (1989) argues, much of the credit often given to Dutch national planning should be shared with (active) local land policy which has helped implementing many plans. Second, 'land policy' in the Netherlands, is generally reserved for those (re)developments that lead to *urban* land-use functions such as housing,

---

[1] As of 2024 this is the physical environment plan as part of the new Environment and Planning Act (Omgevingswet). To prevent confusion and because the effects of the Environmental Law still have to be assessed, we stick to the old terminology.

Fig. 1 *Rijnenburg* in Utrecht, the Netherlands (map: Kleiner, Jehling 2024, Aerial Imagery Provider © Esri, Beeldmateriaal.nl, Maxar, Microsoft)

retail, infrastructure, amenities, and offices.[2] However, no fundamental objection exists to stretching the definition to non-urban uses, such as nature and landscape development.

1.1 Active Land Policy

For many decades, land policy was almost synonymous with what is referred to as *active* land policy, which can be defined as the behavioural pattern aimed at the development of land for urban purposes, *whereby public actors (e.g. municipalities) buy*

---

[2] See for instance a 2023 government letter about the 'modernisation of land policy': Modernisering grondbeleid moet gebiedsontwikkeling versnellen en financieel haalbaar maken (Rijksoverheid.nl).

land themselves, prepare it for development, and sell it under certain conditions to those who take care of the real estate development. The tradition of active land policy finds its origin—although it was not entirely new at the time—in the late nineteenth century, when unbridled housing development led to poor and unhealthy housing, and haphazard urban extension. These conditions were a major force behind the (first) 1901 Housing Act. Although not formally part of it, the Housing Act did provide the impetus for many municipalities to set up a land company (*grondbedrijf*) that would be involved in purchasing, developing, and selling building sites. After all, municipalities were obliged to be more actively involved in public housing and, therefore, with the land underneath it. Moreover, the Housing Act ensured that expropriation 'in the interests of public housing' became possible (Kruijt et al., 1990, p. 97). In 1931, there were already 110 specialized municipal land companies (Ten Have, 2021).

During post-WWII reconstruction, the practice of active land policy became (even) more widespread. Furthermore, the central government became more actively involved as subsidizer of housing and regulator of land prices. A 'golden triangle' of housing development evolved: housing associations built and managed the (social) houses, central government provided the subsidies, and municipalities provided the land at fixed price. The central government limited land prices as it did not want subsidies to leak away unnecessarily to local authorities. This system of sub-market house and rental prices, and limited land prices, virtually led to a municipal land monopoly, which made implementing housing and spatial policies relatively easy (Buitelaar & De Kam, 2012). Moreover, next to the 'golden triangle' a group of municipalities owned a large developer called Building Fund (*Bouwfonds*), which typically also was given the opportunity, often behind the scenes, to acquire land that was to be developed.

1.2 Facilitative Land Policy

Things started to change during the 1990s. By then, the composition of the housing production changed from mainly social housing to mainly owner-occupied homes at market rates. This also aroused the interest of private parties in land development,— just like at the end of the nineteenth century. In addition, in the mid-1990s, housing associations became financially independent and self-supporting, housing subsidies were abolished, and municipalities (as a result) no longer always saw a reason to provide land to housing associations at a fixed low price (Buitelaar, 2010). Building Fund was privatised in 2000 and became by far the largest private developer in the Netherlands under the name of Bouwfonds Property Development (BPD). Although many houses have been developed on municipal land since then, the municipal land monopoly has made way for a much greater variety of landowners, including property developers and investors. The global financial and economic crisis of 2008 has reinforced this heterogeneous ownership landscape as local authorities become more reluctant to buy land.

Table 1 Different instruments for active and facilitative land policy

	Active land policy	Facilitative land policy
Private law	Land assembly/sales	Development agreement
Public law	Pre-emption rights Compulsory purchase	Development plan Inclusionary zoning/housing

When private landowners are willing and able to develop within the boundaries of a land-use plan, municipalities must rely on their planning powers (under public law) and pursue a facilitative land policy.[3] Henceforth, this type of land policy has increased in prominence since the turn of the century.

However, because of the reliance on active land policy for many decades, public law for facilitative land policy had been neglected and, therefore, remained ill-equipped. When the municipal land monopoly started to crumble, it appeared to the legislature that the planning powers under public law had to be strengthened. This is what happened with the enactment of the 2008 Spatial Planning Act. Among other things, this Act provided possibilities to designate land for social housing (and other housing types)—'inclusionary housing'—and it included the coercion of the recovery of costs of public goods and services from private developers. Before the Act, these options could only be imposed if local authorities owned the land (Buitelaar & De Kam, 2012).

1.3   Instruments for an Active and Facilitative Land Policy

In active land policy, local authorities usually perform like any other actor in the land market, by using private law (i.e. the right to ownership) to acquire land. This can be referred to as 'amicable land acquisition'. When a plot of land needed for implementing a land-use plan is in private hands, and the respective private owner does not want to sell it, the municipality may consider the compulsory purchase of the land (*onteigening*). Expropriation can only occur when the landowner is unwilling and unable to implement the plan. It is not often used in practice, as it is a lengthy process, and a full compensation of income and capital loss needs to be paid (Korthals Altes, 2014). However, it provides a 'shadow' under which amicable land acquisition occurs and arguably facilitates and eases that (see Table 1).

Land assembly may also be assisted by the designation of a so-called pre-emption right (*voorkeursrecht*) to a site. Based on a strategic spatial vision that promotes urban (re)development, a municipality may impose a right that contains 'the right to first refusal' for three years, possibly extended by another three years. This means the

---

[3] 'Facilitative land policy' is a synonym for 'passive land policy'. In Dutch discourse 'facilitative' has a more positive connotation than 'passive' and is commonly used. 'Facilitative' is arguably also a more apt term since municipalities are never completely 'passive' in land policy.

landowner must always offer the land he/she wishes to sell to the municipality within this period. The intent is to counter speculation and to keep land prices in check.

The land-use plan is used to impose specific land-use and building conditions. Since the Spatial Planning Act 2008 that has also included the possibility of 'inclusionary housing' and the possibility for municipalities to recover costs (*kostenverhaal*) from developers via a private development agreement (*anterieure overeenkomst*). When developers and municipalities cannot agree on the conditions, municipalities have a second option to recover costs: a public and enforceable development plan (*exploitatieplan*).

2 Resources

How formalised the planning system in the Netherlands may be, the use of land policy instruments is always embedded *in* and results *from* informal practices. We use the case of the polder *Rijnenburg* to illustrate this. *Rijnenburg* is considered a representative case as the issues that play out here are illustrative for plan-making and land assembly in Dutch planning practice (and, arguably, elsewhere).

Rijnenburg (covering 1000 hectares) is a greenfield site located southwest of Utrecht, the Netherlands (Fig. 1). In the early 1990s, the national government, in close collaboration with local authorities, worked on a national policy document for spatial planning called *Vinex* (Ministry of Housing, Spatial Planning and the Environment, 1991). The policy document included development directions for housing in and close to major urban areas, among others, in and around Utrecht. For Utrecht, there were two possible housing locations: *Rijnenburg* and *Leidsche Rijn* (an area north of *Rijnenburg*, separated by motorway A12). Although *Leidsche Rijn* was chosen (now the largest *Vinex* area with over 30,000 houses), *Rijnenburg* was considered the next large development location once *Leidsche Rijn* would be completed (then foreseen in 2015). This sparked the interest of private developers, who started assembling land and establishing option rights[4] on land owned by farmers in *Rijnenburg*. More prospects of future land development followed in 2004 (with the national policy *Nota Ruimte*), when *Rijnenburg* was no longer considered to be part of the Green Heart, a sparsely populated area with a rural character where a building ban applied (Ministry of Housing, Spatial Planning and the Environment, 2004). Following this lift of the national 'ban', the provincial government designated the land for housing in their strategic policy (*streekplan*). Nevertheless, despite the intentions to develop housing in the future, as well as many attempts to change the course of land development, *Rijnenburg* remained an agricultural landscape.

Housing development has arguably been a most prominent proposal for the future of *Rijnenburg* since the early 1990s. Different plans for housing have been presented by a variety of stakeholders. For instance, in 2005 developers proposed constructing 10,000 to 14,000 luxury and suburban dwellings, and in 2009 the

---

[4] This is an option to buy the land when the owner wants to sell it.

City of Utrecht presented a vision with 7,000 homes (Gemeente Utrecht, 2009). Then, the global financial crisis of 2008 and its aftermath heavily disturbed the decision-making process regarding *Rijnenburg*. Before plans became concrete, the plan-making process came to a standstill. Housing development came back on the agenda in the mid to late 2010s, when housing prices had started rising again—they roughly doubled between 2013 and 2022—and the business case for development had become more favourable due to low interest rates, demographic and economic growth, and a shortage of housing supply (Jonkman et al., 2022).

Throughout the 2010s, different discourses than housing development had gained traction though, putting the future of *Rijnenburg* in different perspectives and criticising the idea of housing development. First of all, it had become apparent that urban land was becoming an increasingly scarce resource. The national focus in land policy shifted from greenfield development to inner-city densification. This made the idea of housing development in *Rijnenburg* rather contentious, as the polder was a typical stage for greenfield development.

Second, new activities and land uses had started to compete for prominence over the land in *Rijnenburg*. Tensions came up between different claims for land use—not only in *Rijnenburg*, but in the Netherlands in general (Vejchodská et al., 2022). For instance, in light of the 2015 Paris Agreement, the cause of climate mitigation became a contender for future land use in *Rijnenburg*. As the urgency and political importance of climate mitigation grew during the 2010s, land-based wind turbines were increasingly seen as a solution to mitigate climate change, and locations were sought to locate wind turbines. *Rijnenburg* emerged as a promising location for generating wind energy (Gemeente Utrecht, 2013, 2016), and plan-making for a temporary 'energy landscape' with solar panels and wind turbines started, resulting in a detailed vision in 2020 (Gemeente Utrecht, 2020).

Also, climate *adaptation* measures received increased attention. For instance, the regional water authority started expressing concerns about the potential development of houses in such a low-lying polder as *Rijnenburg*, large parts of which would be flooded in case of extreme rainfall (HDSR, 2022). In general, climate-proof development gained prominence in planning policy and practice during the late 2010s and early 2020s. Ultimately, the central government decided in 2022 to make water and soil conditions leading aspects in future land development (*water- en bodemsturend*) (Rijksoverheid, 2022).

3   Actors

The case of *Rijnenburg* is characterised by tensions between different agendas and claims that are articulated, promoted, and pushed by stakeholders and shareholders in the area, and the different degrees of (relational) power the respective actors can exert. Moreover, actors' agendas and claims have changed over time. In general, three (groups of) actors have been particularly relevant.

First, real estate developers became crucial actors when they started assembling land from farmers in *Rijnenburg* in the early 1990s, ahead of potential land acquisitions by the local authority that was primarily occupied with the large development of *Leidsche Rijn*. In the years between 1993 and 2003, developers acquired a total of 735 hectares in the area, and land acquisition peaked in 2004 with 175 hectares in one year (Buitelaar et al., 2025). As such, land in *Rijnenburg* has largely remained privately owned, with real estate developers and housing associations as the largest shareholders; they own over 70 per cent of the land that is potentially suitable for development (Fig. 3). These actors have always advocated housing development in *Rijnenburg*, yet in different ways, at different levels of intensity, and with several changes of plans throughout the years. For instance, soon after 2004, a small consortium of large developers first presented their ambitions to the municipality. More recently, all developers owning significant amounts of land in the area teamed up in a large consortium and intensified their collaboration to influence the decision-making process. The more recent plans of this consortium were to propose 25,000 to 35,000 houses.

Second, the City of Utrecht has been the regulator of the land. The municipality presented its first detailed vision for *Rijnenburg* in 2009, which included a low-density residential area (City of Utrecht, 2009). Developers owning land in *Rijnenburg* were not convinced by these plans, which allegedly lacked financial feasibility. The City of Utrecht invited developers to propose alternatives, which illustrates the

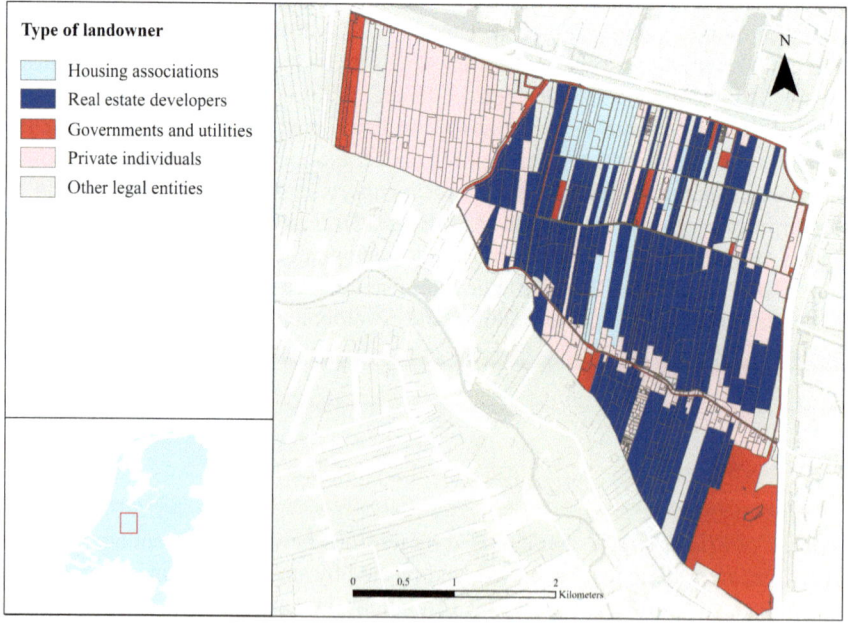

Fig. 3 Landownership in *Rijnenburg* (*Source*: The Netherlands Cadastre, Land Registry and Mapping Agency)

generally constructive interactions between the local authorities and *Rijnenburg*'s private landowners at the time. Consultations between the municipality's executives and public managers on the one hand and developers on the other were common, perhaps partly because the interests between the different sectors did not conflict fundamentally: both the City of Utrecht and the developers saw *Rijnenburg* as a future housing location.

A third and final relevant actor came in later and changed the playing field significantly: the rather unconventional vision of *Rijnenburg* as an energy landscape corresponded with the ambitions of Rijne Energie, a local energy collective pursuing the construction of at least four wind turbines in *Rijnenburg*. If *Rijnenburg* were to become an energy landscape, housing development would no longer be an option, at least for several decades. Rijne Energie's entry to the arena after 2015 put the plans of the developers at stake, so the latter responded by increasing their activity level. Two large developers—BPD and AM—founded a large consortium of landowners, including other developers, investors and housing associations (13 partners in total). The consortium partners had the same interest (i.e. developing housing) and framed *Rijnenburg* as the most appropriate area for the City of Utrecht to develop housing. Toward the city elections of 2022, the consortium actively engaged in the public debate, launching a campaign titled 'Vote for *Rijnenburg*'. This campaign and a series of other activities put housing development on the political agenda again, thereby actively opposing the position of the City of Utrecht. A series of events ensued that illustrated the deteriorated relationship between developers and the City of Utrecht—a relationship that had generally been positive and constructive since the future of *Rijnenburg* had been a topic of discussion.

Other government actors than the City of Utrecht have mainly remained in the background. The visions of the provincial government and the municipality have been similar, and the central government—despite a motion passing in parliament to dictate housing development in *Rijnenburg*—has not shown much interest in interfering with the decisions of local authorities. The region's water authority is an exception, as it explicitly articulated the importance of climate adaptive building and flood protection towards local media and the municipality. Regarding private actors, most of the local residents cannot be considered shareholders anymore, as they sold their land or put it under option. Some residents, mainly from surrounding areas, have been opposing the construction of wind turbines and founded an organisation called 'Neighbours of *Rijnenburg*' (*Buren van Rijnenburg*).

The dynamics observed in actors' agendas, positions, and claims demonstrate increased societal challenges and conflicts to be dealt with in politics and policies for land development. Not only have *more actors* become involved in such planning questions (i.e. plurality); we have also witnessed new and different *types of actors* coming in (i.e. pluriformity). Consequently, it has become more difficult to reach consensus about future developments.

4 Institutions

Property rights in land are critical institutions for understanding the course of land-use planning and land development (Needham, 2006). Inspired by the French Revolution and the *Code Civil*, the Netherlands has a legal system with a strong position of private property rights (Van den Bergh, 1988). Those rights give large control over both the use of and the income from a good such as land. Therefore, obtaining those rights is very important for being able to exert economic and social power (Claassen, 2023). However, property rights are never absolute; they are restricted, among other things, by rules from public law, including the rules in a land-use plan. Moreover, property can be expropriated by the state; in the case of the Netherlands the option for that is relatively far-reaching, and so the financial compensation (Alterman, 2010).

Landowners exert power in the regulatory process towards a land-use plan because they *can* and *want* to. First, there is room for manoeuvring because land-use regulation is generally susceptible to change, allowing developers to get a foot in the door. This room for manoeuvring is not only the case in common-law systems which are known for their dynamic case-by-case approach, such as the British system (Corkindale, 1999). There is also an inclination to regulate in detail, on a case-by-case basis, in statutory systems. A land-use plan is not a general 'urban code', but a site-specific regime that quickly becomes obsolete and requires revision to accommodate social change (Booth, 1996; Moroni, 2007). In the Netherlands, a country with a statutory legal system, the majority of land-use plans are made, amended, or revised altogether in response to development initiatives that do not fit within the legal regime that existed before revision; in contrast to the popular belief, the Dutch planning system is development-led rather than plan-led (Buitelaar et al., 2011).

In the case of *Rijnenburg*, room for manoeuvring was created—and continues to be created—as several governments have been more or less openly discussing the future use of the area. These open discussions, particularly in the 1990s made the area a prospect, and as such created the room for developers to assemble rights on land (economic ownership, legal ownership, or option rights) in *Rijnenburg*.

While regulatory practice creates *room* for developers to get involved, for commercial parties 'land rents' create the *desire* to use that room. Land-use regulation determines what uses are allowed on a plot of land, and the land use determines its value (Van Dijk & Van der Vlist, 2015). The difference between the value of the produce of the land and the cost of making that happen is commonly referred to as the 'residual value' of land (Evans 2004). The difference between the residual value and the existing use value is what economists call 'land rent' (Oxley, 2004). Rents are worth capturing, and doing so turns developers into 'rent-seekers'. Particularly in the case of housing development on greenfield land, the value uplift can be substantial since the residual value of land for housing is equivalent to multiple times the agricultural value of that same land (Adams & Watkins, 2002). Consequently, the stakes of rezoning land from agriculture to housing become higher.

Rijnenburg is the only remaining greenfield site within the boundaries of the City of Utrecht, the fastest-growing city in the Netherlands with one of the tightest housing markets. As a result, potential land rents could be enormous. It incentivises developers and other actors to become landowners and subsequently engage in strategies of institutional entrepreneurship aimed at land-use plan revision in a way that captures those potential land rents (Buitelaar et al., 2025).

4.1 Land Development and Plan-Making

Once the local authority consents to the housing development in *Rijnenburg*, the plan-making and land-development stage will commence. Given the land-ownership structure of several more prominent landowners, and the Dutch land development tradition (e.g. Bregman et al., 2018), there are roughly two potential modes of development.

First, the landowners, including the municipality, could pool their land and transfer it to a new public–private legal entity, a joint venture, which can undertake the development, acquire the remaining necessary plots, and deliberate with the municipality and other stakeholders about the plan. At completion, each shareholder gets the same share of the entered value. After the financial-economic crisis of 2008–2013, this mode became less popular as many of these joint ventures ran into financial troubles, leading to private developers withdrawing, leaving behind the public actors and significant public debts.

Second, the landowners may decide to transfer all land to the municipality, which then takes care of land development (e.g. preparation, remediation of contamination), after which it transfers back the land and issues building rights to the actors that handed in the land in the first place. This mode is referred to as the building claim model (*bouwclaimmodel*). It gives the municipality greater control over the processes and reduces the financial risks for developers. Since the 1990s, this mode has been dominant in housing production (Groetelaers, 2004).

Whichever mode the shareholders pick, they would be closely involved in revising the land-use plan. Furthermore, in any case they would be required to cover of the costs of the land preparation and the plan-making (*kostenverhaal*). This will be settled during the land transfer or by fee when issuing the building permit.

5 Reflection

In this chapter, we have tried to explain and illustrate the workings of Dutch land policy. Dutch spatial planning was traditionally internationally praised for its effectiveness, and so is land policy for its role in the implementation (Needham, 1989). However, not only has the capacity of spatial planning and land policy eroded over the last decades, but actual practice has always been much more pragmatic and fuzzier than the perfect image that is sometimes portrayed. Put differently, it was never a planner's paradise to begin with but an ambiguous utopia at best.

The case of *Rijnenburg* shows the confrontation and trade-off between different interests that play out in the polycrisis. What becomes clear is that in the system of facilitative land policy short-term interests, such as reducing the housing shortage, tend to prevail over long-term interests, such as climate mitigation and adaptation. This is sometimes referred to as 'tyranny of the now' (Krznaric, 2021). Resisting pressure from powerful stakeholders with financial 'skin in the game' is extremely difficult for local politicians and policymakers, if not impossible. In particular because we observe that developers buy into the discourse of the housing crisis and frame building on greenfield sites as a relatively quick and easy solution. These pressures are hard to withstand in electoral cycles where voters ask for solutions for tangible problems. In *Rijnenburg*, developers even launched a campaign in the municipal elections, making them become (co-)producers of the rules of the game, so called institutional entrepreneurs (Buitelaar et al., 2025).

They can become institutional entrepreneurs because of the nature of development control in the Netherlands. Although the Netherlands has a plan-led planning system on paper, practice shows that plans tend to get drafted in response to development initiatives and do not provide any upfront guidance. Practice is development-led rather than plan-led (Buitelaar et al., 2011). There is a difference between *de jure* and de facto, between formal and informal. So, the way the system is used gives room for institutional entrepreneurship by land developers. Recently, the national government has become more aware of the pitfalls of this approach, including the risk of speculation, and has developed a proposal to modernise land policy (*Modernisering Grondbeleid*, Ministry of the Interior and Kingdom Relations, 2023).

Not only private developers are 'entrepreneurial', public authorities too can act along the lines of private entrepreneurs. In the Netherlands, public authorities do not only regulate the market, they both play the game and make the rules of the game (Buitelaar et al., 2022). Where the 'rule of law' is aimed at a clear distinction between the public and the private (Moroni, 2007), in Dutch practice we see hybridization. Local authorities often engage in all sorts of public–private partnerships and use private law to pursue their (public) goals. While this approach has proved relatively successful to achieve short- or medium-term goals, like housing development, it proves much more challenging for long-term goals, because business models and land account instruments do not consider long-term costs and revenues, for instance relate to climate adaptation (Buitelaar & Pelzer, 2024). In light of the polycrisis we started this chapter with, considering the long-term in both plans and actual practices is a necessary condition for the Netherlands to become a planner's paradise of the twenty-first century.

References

Adams, D., & Watkins, C. (2002). *Greenfields, Brownfields and Housing Development*. Blackwell.

Booth, P. (1996). Controlling development. Certainty in discretion in Europe, the USA and Hong Kong. UCL Press.

Bregman, A., Karens, J., Buitelaar, E., & De Zeeuw, F. (2018). Gebiedsontwikkeling in een nieuwe realiteit. Instituut voor Bouwrecht.

Buitelaar, E. (2010). Cracks in the myth: Challenges to land policy in the Netherlands. *Journal of Economic and Social Geography, 101*(3), 349–356.

Buitelaar, E., & De Kam, G. (2012). The emergence of inclusionary housing: Continuity and change in the provision of land for social housing in the Netherlands. *Housing, Theory and Society, 29*(1), 56–74. https://doi.org/10.1080/14036096.2011.592214

Buitelaar, E., Galle, M., & Sorel, N. (2011). Plan-led planning systems in development-led practices: An empirical analysis into the (lack of) institutionalisation of planning law. *Environment and Planning a: Economy and Space, 43*(4), 928–941. https://doi.org/10.1068/a43400

Buitelaar, E., Van den Hurk, M., Nozeman, E., & Oude Veldhuis, C. (2022). Public entrepreneurialism in private land markets. Contracting dilemmas around selling Amsterdam's major prison. *Planning Theory and Practice, 23*(2), 248–264. https://doi.org/10.1080/14649357.2022.203 4923

Buitelaar, E., & Pelzer, P. (2024). Short-term gains and long-term losses. Climate adaptation trajectories facing land accounting institutions. *Paper presented at the Annual AESOP conference*, Paris, 8–11 July 2024.

Buitelaar, E., Lebbing, J., Pelzer, P., Van den Hurk, M., & Van Karnenbeek, L. (2025). Regulate or be regulated: the institutional entrepreneurship of landowners. Planning Theory & Practice. https://doi.org/10.1080/14649357.2025.2456865

Campbell, H. (2006). Is the issue of climate change too big for spatial planning? *Planning Theory & Practice, 7*(2), 201–230. https://doi.org/10.1080/14649350600681875

Claassen, R. (forthcoming). What we own and what we owe. Property as a power for the common good.

City of Utrecht. (2009). Wonen in het landschap: Structuurvisie *Rijnenburg*. Retrieved from: https://www.commissiemer.nl/docs/mer/p23/p2331/2331-071structuurvisie.pdf

City of Utrecht. (2013). Structuurvisie Windmolens Lage Weide. Retrieved September 6, 2024, from: https://utrecht.bestuurlijkeinformatie.nl/Document/View/949fc100-488c-4849-818b-952 e8c9ab4e3

City of Utrecht. (2016). M107 Bouwpauze in Rijnenburg. Retrieved September 6, 2024, from: https://utrecht.bestuurlijkeinformatie.nl/Reports/Document/24f0c37b-e1f5-46da-9677-03dc810fd420?documentId=d35355c2-9865-403a-aeb0-2e6dd8e228c5

City of Utrecht. (2020). Visie Energielandschap Rijnenburg en Reijerscop. Retrieved September 6, 2024, from: https://www.commissiemer.nl/projectdocumenten/00009472.pdf

Corkindale, J. (1999). *Reforming land-use planning*. Institute of Economic Affairs.

Groetelaers, D. A. (2004). *Instrumentarium locatieontwikkeling: Sturingsmogelijkheden voor gemeenten in een veranderde marktsituatie*. Delft University Press.

HDSR. (2022, February 24). Bouwen in Rijnenburg: nee, tenzij klimaat- en bodemdalingsbestendig. Retrieved September 6, 2024, from: https://www.hdsr.nl/zoeken/@143668/bouwen-rijnenburg/

Jonkman, A., Meijer, R., & Hartmann, T. (2022). Land for housing: Quantitative targets and qualitative ambitions in Dutch housing development. *Land Use Policy, 114*, 105957.

Korthals Altes, W. K. (2014). Taking planning seriously: Compulsory purchase for urban planning in the Netherlands. *Cities, 41*(Part A), pp. 71–80.

Kruijt, B., Needham, B., & Spit, T. (1990). Economische grondslagen van grondbeleid. SBV.

Krznaric, R. (2021). *The good ancestor: How to think long term in a short-term world*. Ebury.

Lucassen, M. (2002, December 18). *Utrecht houdt hoog bouwtempo aan*. AD Utrechts Nieuwsblad, p. 1.

Ministry of the Interior and Kingdom Relations (2023). *Kamerbrief modernisering van het grondbeleid*. The Hague.

Ministry of Housing, Spatial Planning and the Environment (1991). *Vierde nota ruimtelijke ordening extra*. The Hague.

Moroni, S. (2007). Planning, liberty and the rule of law. *Planning Theory, 6*(2), 146–163. https://doi.org/10.1177/1473095207077586

Needham, B. (1989). Planning and the shape of the Netherlands through foreign eyes: But do appearances deceive. *Built Environment, 15*(1), 11–16.

Needham, B. (2006). *Planning,* law and economics: the rules we make for using land. Routledge.

Oxley, M. (2004). *Economics, planning, and housing.* SAGE.

Rijksoverheid. (2022, November 25). Kabinet maakt water en bodem sturend bij ruimtelijke keuzes. Retrieved September 6, 2024, from: https://www.rijksoverheid.nl/actueel/nieuws/2022/11/25/kabinet-maakt-water-en-bodem-sturend-bij-ruimtelijke-keuzes

Ten Have, F. (2021). Waarom verliezen in het verleden geen beperking hoeven te zijn voor grondbeleid in het heden. Gebiedsontwikkeling.nu, 19 October. Retrieved September 6, 2024, from: https://www.gebiedsontwikkeling.nu/artikelen/laat-verliezen-uit-het-verleden-geen-beperking-zijn-voor-actief-gemeentelijk-grondbeleid-in-het-heden/

Van Buuren, P. J. J., Nijmeijer, A. G. A., & Robbe, J. (2017). *Hoofdlijnen ruimtelijk bestuursrecht.* Wolters Kluwer.

Van den Bergh, G. C. J. J. (1988). *Eigendom.* Kluwer.

Van Dijk, T., & Van der Vlist, A. (2015). On the interaction between landownership and regional designs for land development. *Urban Studies, 52*(10), 1899–1914. https://doi.org/10.1177/0042098014544764

Vejchodská, E., Shahab, S., & Hartmann, T. (2022). Revisiting the purpose of land policy: Efficiency and equity. *Journal of Planning Literature, 37*(4), 575–588. https://doi.org/10.1177/08854122221112667

Edwin Buitelaar is a Full Professor of Land and Real Estate Development at Utrecht University, and Research Director of the Department of Human Geography and Spatial Planning of that same university. He is also a research associate at PBL Netherlands Environmental Assessment Agency. His researches focuses on the land and real estate development, institutions, and justice.

Martijn van den Hurk is an Assistant Professor of Spatial Planning in the Department of Human Geography and Spatial Planning at Utrecht University. His scholarship is aimed at understanding the institutional arrangements that underlie transformations in urban infrastructures and city building. He is particularly interested in questions revolving around contracts and (public) accountability in public–private partnerships and urban redevelopment projects. Van den Hurk completed a PhD in Political Science at the University of Antwerp.

Jasper Lebbing is policy advisor at the Dutch municipality of Bunnik and he was—at the moment of the research—a junior researcher at the Department of Human Geography and Spatial Planning at Utrecht University. During the research for this book chapter, he completed his master degree in spatial planning with honors (cum laude). His research interests lie in the interaction between actors and planning institutions, and local land issues. To bridge the gap between theory and practice, he is currently working as a policy advisor for a local government.

Peter Pelzer is a Full Professor of Spatial Planning and Strategy in the Department of Urbanism at TU Delft. His research centers around imagination, longtermism and transitions related to the physical environment. He was part of an expert committee of the Council for the Environment and Infrastructure that advised the Dutch government on climate adaptation and spatial planning. In 2024, he was the first planning scholar to become the Dutch academic teacher of the year.

Lilian van Karnenbeek is a postdoctoral researcher at the Institute of Jurisprudence, Constitutional and Administrative Law at Utrecht University. She participates in the Utrecht University Centre for Water, Oceans and Sustainability Law research program. Lilian holds a PhD in Urban Planning. She researches European and national laws and policies on climate change adaptation, particularly on floods, water nuisance, and drought.

Reflections

12 Chapters of Land

Benjamin Davy†

Abstract The comment discusses the selections of "land" by 12 country case studies. Case studies that examine specific sites allow to take a closer look, but make it difficult to distinguish between typical and random aspects of a case. One of the perplexing aspects of land policy are the diverse meanings of land. Possible meanings include land as an economic good with an exchange value, land as an environmental resource with an existence value, land as a functionality with a use value, or land as a source of spatial power with a territorial value. In the 12 chapters, a recurring theme on land is the re-use of brownfields or the "densification" of land that in the minds of planners and other policymakers is underused. Also, most of the 12 country reports focus on restricted land uses and private property rights. The case study approach renders most interesting results with respect to the meaning of land and its impact on land policy goals. It illustrates that political conversations about land are not merely emp-ty talk. Language determines, for example, whether land is predominantly considered as an economic good that can be sold and bought or as an environmental resource that must be protected from reckless exploitation.

1 Selections of "Land"

Each of the 12 chapters describes "land" as a resource that has been selected as the object of each country case study. Usually, this land is proposed for development in an urban context, often in the town centre but occasionally in the urban fringes. The plot size ranges from under one hectare to 1000 hectares, yet often the plot size remains unspecified. The current uses of the land include "empty", "wild", "former industrial", and "office space". Table 1 summarises the examples of "land" examined by the country reports.

Academic literature on land policy often considers land in broad terms, for example, as urban or rural land, as building land or open space, as farmland or forest. A broad view on land necessarily generalises complex details on the ground. Case

B. Davy† (✉)
University of Johannesburg, Johannesburg, South Africa

© The Author(s) 2025
T. Hartmann et al. (eds.), *Land Policies in Europe*,
https://doi.org/10.1007/978-3-031-83725-8_14

Table 1 "Land" examined in the cases

Cases from	Selection of "land"
Austria	A small hamlet (Kaisers, Sölden), size not specified
Belgium	0,63 hectares of wooded area (Akkerstraat) and 20 hectares of forest linked to a castle park (Slotendries) in the City of Ghent
Czechia	1,15 hectares of wild green area (K Pazderkám, Prague)
England	Seven non-adjacent sites in the Wirral Green Belt, size not specified
Finland	33 hectares of vacant former industrial site close to the center of Joensuu
France	Two development zones in Rue Romain Rolland (in Bouleurs, Greater Paris metropolitan region), size not specified
Germany	0,37 hectares of "almost empty plot of land" (Burgwall 21, Dortmund)
Norway	200 hectares of densification area (Mindemyren, Bergen)
Poland	Empark Mokotów Office Business Park, 107,000 m^2 of office space, plot size not specified
Sweden	Former industrial zone Norra Djurgårdsstaden, size not specified
Switzerland	Greenfield area (Viererfeld, Bern), size not specified
The Netherlands	1000 hectares of greenfield site (Rijnenburg, Utrecht)

studies that examine specific sites allow their authors to take a closer look. Single case studies make it difficult, however, to distinguish between typical and random aspects of a case. Will building a rockfall protection dam (Austrian case) always facilitate land readjustment or is this only true for Kaisers, Sölden? Is climate adaptation a regular obstruction to greenfield developments (Dutch case) or a random problem that emerged in the *Rijnenburg* case? Two cases deal with undeveloped land in situations when negotiation skills will determine the outcome, not land policy (Czech and German case).

The following reflection on "land" considered by the 12 chapters focuses on the plural meanings of land, on the relationship between restricted and shared uses of land, and on the link between "land" and the objectives of land policy.

2   Plural Meanings of Land

One of the perplexing aspects of land policy are the diverse meanings ascribed to land (Davy, 2012). Possible meanings include land as an economic good with an exchange value, land as an environmental resource with an existence value, land as a functionality with a use value, or land as a source of spatial power with a territorial value. The plural meanings of land can be illustrated by typical views of different stakeholders and actors:

- Developers consider land as an economic commodity that is subject to competitive pricing and bought and sold on land markets.
- Ecologists address land as a scarce environmental resource that combines biodiversity, soil qualities, plants, animals, and water in complex ways.
- Planners think of land often in terms of functionalities: What are the purposes for which land is currently used and should be used in the future? Land use plans, after all, are instruments of assigning specific land uses to urban land.
- Lawyers think of land as properties that render many rights and some duties to the landowners. In this sense, land are bundles of rights assigned to surveyed and registered plots of land.

In practice, policymakers often find it surprisingly easy to ignore the meanings of land which they do not need for creating a specific land policy. In many countries, the result of pragmatic policymaking are clusters of land-related policies that often fit quite well with each other—simply because ignorance and negligence are proven strategies to deal with polyrationality. The sociologist Georg Simmel labelled the phenomenon as "mental condom" (Simmel, 1903) which protects urban dwellers from an overflow of information and emotions. From a theoretical perspective, however, mental condoms prevent researchers from gaining insights into the social construction of land and its plural meanings. In order to avoid the pitfalls of monorationality, planners and other policymakers must engage in "expectation management" (Hartmann, 2012).

In the 12 chapters, a recurring theme on land is the re-use of brownfields or the "densification" of land that in the minds of planners and other policymakers is underused. The narrative often starts with conflicting plural values of land that emerge from simultaneously wishing to create more housing and prevent more open space from being converted into building land. If, for the sake of the argument, such a conflict is reduced to two parties (the builders and the sentinels), these parties will often not even agree on what the problem is. The builders think of the land as an important input into their proposed development. Builders prefer certainty because every surprise changes the calculated price of the development at the cost of the builders' profits. Also, builders prefer the most extensive use of the land: each additional square meter of residential or retail floorspace translates into additional income. The sentinels could not care less about maximising floorspace. So much land already has been converted into single family units, office towers, parking lots and ecologically impoverished townscapes. Sentinels assert that the biodiversity of land has matured in peace for decades, nay centuries, and provides nourishment for wild plants and feral animals.

What can planners do to address the aspirations and anxieties of builders and sentinels? Surely, planners cannot decide which of the two parties is right or wrong. Planners rather develop plans that acknowledge the concerns of builders and sentinels. Re-using former industrial land (as seen in the cases from Finland, Poland and Sweden) can help provide a resource for the builders without irritating the sentinels. Another popular strategy is the more intense use of land already designated as building land, often called densification (as seen in the cases in Belgium,

Germany, Norway, Switzerland). In both cases, the development of fresh projects is made possible without converting more open space into building land. Brownfield re-development as well as densification—sometimes naively—assume that the quantitative neutrality equals the qualitative neutrality of land uses. The assumption is based on a shallow concept of land reduced to its two-dimensional representation in a cadastral map. Thematic maps are paradigmatic examples of Simmel's "mental condom", and once the meaning of land is reduced to a simple cartographic representation, planners and other policymakers feel free to invent whatever meaning of land they desire. The planners' illusion of having the power to define spaces as "empty" (German case) is an expression of territoriality:

> Territoriality in fact helps create the idea of a socially *emptiable place*. Take the parcel of vacant land in the city. It is describable as an empty lot, though it is not physically empty for there may be grass and soil on it. It is emptiable because it is devoid of socially or economically artifacts or things that were intended to be controlled. (Sack, 1986: 33–34).

Re-use and densification of urban land acknowledges the diverse meanings of land as defined by builders and sentinels. This does not mean, however, that Simmel's "mental condom" works for every stakeholder. In fact, re-use or densification efforts frequently meet resistance from local residents, who do not perceive an underused plot of land as "vacant" and who believe that a more intense use of the land will be detrimental to their neighbourhood.

3 Restricted and Shared Uses of Land

All land uses are either restricted to proprietors (landowners) or shared by members of a use community. Restricted uses are regulated by private property, shared uses by common property relations. Paradigmatically, a private home is designated for the restricted use by an occupant and their family, a public street enables its shared use by the general public. Well-balanced spatial patterns always combine restricted and shared uses, private as well as common property relations. Land policy must account for this combination. Neither a focus on private properties (typical of a private law perspective) nor a one-sided fondness for public spaces (typical of urban sociology) helps understand and shape spatial patterns (Davy, 2012: 231–233).

Most of the 12 country reports focus on restricted land uses. Perhaps this focus can be attributed to the strong protection of private property rights in many of the legal systems (as seen, for example, in the cases in Austria, Belgium, and Poland). If a country is described as "landowners' paradise" (Belgian case) or a Warsaw district is called "Mordor", it stands to reason that in these countries the interest for the shared uses of land is low. Notable exceptions are the mentioning of shared land uses in the case of a rockfall protection dam (Austrian case), green belts (English case), public space (Norwegian case), or climate action (Dutch case).

The neglect of the availability of land for shared uses is a popular mistake about how cities are functioning and how most people are using urban spaces. Most urban residents do not stay in their private mansions or apartments the whole day. They travel around in their cars or on their electric bikes, visit doctors, public pools, or museums, meet for business lunches in restaurants, take a stroll in a pedestrian zone, or spent an afternoon in the shopping mall. All of these activities which are part of the urban experience include the shared uses of land. Often, the urban commons do not even have an exchange value (because they belong to the municipality and are not supplied through the land market). The lack of exchange value, however, does not mean at all that urban land for shared uses is economically unimportant. Rather, this land is a prerequisite of a successful and rewarding urban experience.

Several of the country reports perceive "land" in a way that neglects urban commons. Ignoring common property relations creates imbalanced and unproductive spatial patterns. The awareness for the nexus between private and common property is essential for innovative land policy (Davy, 2023).

4   The Meaning of Land and Its Impact on Land Policy Goals

The case study approach renders most interesting results with respect to the meaning of land and its impact on land policy goals. In some of the countries where the land available for settlement is limited by the topography and other natural factors, land policy is always associated with scarcity management (cases from Austria, Belgium, Switzerland, and the Netherlands). In countries with vast open spaces, the viability of land as the location for vulnerable urban uses is more important (cases from Finland, Norway, and Sweden).

The impact of the meaning of land on the goals of land policy is particularly significant in cases when the meaning of land is politically re-defined. The Belgium concept of "land use allocation neutrality" is a stimulating example in case (Belgium case). The country report describes Belgium as "landowners' paradise" because historically, Belgian land use planning and land policy relied heavily on the interests of private landowners. The 2018 strategic plan of the City of Ghent introduced the objective of *ruimteneutraliteit*, i.e., land use allocation neutrality. This objective only makes sense if land is no longer perceived as an individual plot of land that exclusively serves the interests of its owner. If land is conceived as an agglomeration of advantages and disadvantages (often called internal and external benefits and costs), land policy must account for the entirety of effects associated with the use of land (Buitelaar, 2007). The oversizing of building land ceases to affect only the private landowner, but commences to become a growing burden on the public interests. The policy reaction—*betonstop* (concrete stop) or *bouwshift* (construction shift)—has been legitimised by the gradual shift in the way the meaning of land was shaped in public discourse.

The observation about the meaning of land and its impact on land policy goals illustrates that political conversations about land are not merely empty talk. Language determines, for example, whether land is predominantly considered as an economic good that can be sold and bought or as an environmental resource that must be protected from reckless exploitation.

References

Buitelaar, E. (2007). *The cost of land use decisions*. Blackwell.
Davy, B. (2012). *Land policy*. Ashgate.
Davy, B. (2023). Innovative property for innovative land policy: Four normative principles. *Raumforschung und Raumordnung Spatial Research and Planning 81*(6), 648–652. https://doi.org/10.14512/rur.1702
Hartmann, T. (2012). Wicked problems and clumsy solutions: Planning as expectation management. *Planning Theory, 11*(3), 242–256. https://doi.org/10.1177/1473095212440427
Sack, R. D. (1986). *Human territoriality*. Cambridge University Press.
Simmel, G. (1903). Die Großstädte und das Geistesleben. In T Petermann (Ed.), *Die Großstadt. Vorträge und Aufsätze zur Städteausstellung. Jahrbuch der Gehe-Stiftung zu Dresden* (Vol. 9, pp. 185–206). GeheS-Stiftung zu Dresden.

Ben Davy graduated from the University of Wien (Austria) law school. Until his retirement, he was Chair of Land Policy, Land Management, and Municipal Geoinformation at the School of Spatial Planning, TU Dortmund University (Germany). Among Ben's monographs are Essential Injustice (1997) and Land Policy (2012). Ben was Vice President and President of the International Academic Association of Planning, Law, and Property Rights (PLPR) and of the Association of European Schools of Planning (AESOP). Since 2019, Ben has been a visiting professor at the Law Faculty of the University of Johannesburg.

Land Development in Land Policy Processes: Actors and the Dominance of the Pipe-Line Effect

Tejo Spit

Abstract An analysis of differences and similarities of actors involved in land policy throughout Europe is the main topic of this chapter. The analysis shows that despite the differences in context situations that there are many similarities in the behavior of public, as well as private actors. We discovered that during the long running time of land development projects- programs and/or processes all over Europe, the actors involved depend heavily upon each other. This interdependency limits the scope of choices for each actors, which can be considered the main cause for this similarity. On top of that, the sum of all limiting conditions is determining the room to maneuver for actors, which is often sharply decreasing in time. This is what we would like to call the "pipe-line-effect". The result of which is that there are little possibilities for actors to adapt to changing circumstances: all actors suffer more or less under the same 'tyranny of the now'. During development processes all actors try to adapt to changing circumstances. At the same time there is a urgent need for (policy) consistency. Surprisingly, the case-studies show that private parties do better in that respect than public parties.

1 Introduction: Two Types of Actors

Reflecting on the role and activities of actors involved in spatial development processes implies putting them in the sphere of different systems and different contexts. By definition, the actors possess different goals and instruments and face very different challenges. In short, at face value similar actors may appear to show very different behaviour all over Europe. Underneath the surface however, there are a lot of similarities and shared interests. In this paragraph we shall dive deeper into those differences and similarities in the hope to deepen the understanding of the behaviour of those actors involved in spatial development throughout Europe.

In general, two types of actors dominate the land development scenes all over Europe. *Public* and *private* actors fight over the rights to develop sites in all countries.

T. Spit (✉)
The Netherlands Environmental Assessment Agency, The Hague, The Netherlands
e-mail: tejo.spit@pbl.nl

© The Author(s) 2025
T. Hartmann et al. (eds.), *Land Policies in Europe*,
https://doi.org/10.1007/978-3-031-83725-8_15

This fight, a better description would be *struggle,* concentrates on three issues: the quantity (*how much?*), the quality (*what?*) and the time frame (*when?*). This struggle appears to be a 'balancing act' for all partners involved. As long as some sort of balance can be reached, the development continues, but wherever this 'balance' is out of reach, a stand-still will follow. This situation is sometimes referred to as a 'marriage between state and market' (see the case of the U.K.). The next logical question in this struggle between two types of actors is that of dominance. In other words, who dominates this struggle? The case studies do not show a one-dimensional answer to this question. What does this tell us about dominance?

For the sake of consistency, we will limit ourselves to the process of *land development* and not on *land policy*, which is a far wider concept. The reason for that is that most countries differ significantly in their definition of land policy, which complicates a comparison. On the other hand, there is a broad consensus on more narrowly defined concepts of land development. As land development can be considered a crucial part in all land policy processes, this seems to be the most plausible way to organise a systematic evaluation of actors' behaviour in such a variety of circumstances.

In this paragraph, we will argue that although the institutional setting of each actor is an important determinant for its behaviour, the shared limitations between actors in land development processes are most important. Therefore, we will shortly go into the contextual differences, then try to shed some light on the consequences for the behaviour of actors in the development of land. Next, we will dive into our most important conclusion, addressing the question: why do all those seemingly different case-studies all over Europe essentially look similar? We will close this paragraph with some final observations.

2   The Institutional Arrangements: The Context

2.1   *Differences*

Newman and Thornton (1996) identified five so-called *planning families* in Europe which differ fundamentally from each other. Relevant for this study seem the differences between the Napoleonic, German and Eastern European families. The assumption is that these different circumstances have wide implications for the actors involved in development processes. A closer look at the case-studies in Europe shows that this seems to be true for most of the cases. The institutional differences are not only the cause of widely differing starting positions in developing processes, but especially resulting in different sets of instruments with public actors, as well as private actors. These contextual differences seem to cause a lot of differences between the cases. Although in most (or all?) European countries municipalities are in the lead, in some countries other tiers of government play a role in land development processes (e.g. cases from Czechia, France, Norway, Poland). Sometimes they have an important position, resulting in legal complexity in the first instance and sometimes even to direct conflicts (e.g. case from Norway). A first conclusion in this respect must be that this presence at least complicates the already complex development relations between actors. More important however is that their threat to intervene can be perceived as Damocles' sword.

2.2 Similarities

Yet, looking at the behaviour of all actors involved in development processes, it is not the differences that catch the eye, but the similarities for characteristic of all development processes seem to be the mutual dependency between actors, either formally organised in public-private arrangements or otherwise (in formal/informal arrangements). Consequently, all actors show similar behaviour in their strive for profits. Therewith, we will try and shed some light on the issue to what unifies the actors, instead of their differences, as the latter show themselves clearly enough in each case-study.

3 Room to Manoeuvre (Behaviour and Goals)

3.1 More Actors into the Interplay

As the development processes in spatial planning tend to grow more and more complex, the lack of transparency grows as well. This is caused by two developments which are more or less visible in all countries. The first of which is the increase of rules and regulations in spatial planning, especially in the area of environmental concerns and European regulations. The combination of which is self-evident (see e.g. cases from Poland, U.K. and France). Also, by definition, the not quite transparent real estate market (Spit & Zoete, 2016; Witte & Hartmann 2022) contributes to the complexity, although some cases (German or Austrian case) try to increase the transparency by making information publicly accessible. This intransparency of the property market provides room for new actors to gain importance in these processes. These actors mostly consists of consultants and (real estate) brokers. The example from England shows clearly the growing importance of consultants (*consultocracy*), while in other countries (such as the Netherlands and France), where consultants already had a prominent position, the growth is less visible.

3.2 Behaviour and Goals

To many observers, the behaviour of actors in development processes may seem opportunistic (e.g. case from Germany), incomprehensible (e.g. case from U.K.) or even mystic (e.g. France). Yet, a closer observation of their behaviour might reveal that each actor is just trying to optimise their (sometimes changing) goals within ever changing circumstances (market, policy-development on all levels of government and behaviour of colleagues). Against the background of an untransparent property market, it is easy to understand the above-mentioned types of diagnosis, as the behaviour of actors is hard to rationalise.

Differentiation and fragmentation in the arenas of actors increase in those cases in when other tiers of government intervene (e.g. cases from Poland and France), other types of policymaking have large consequences for development (e.g. Belgium case, Finish case) and when private actors (such as landowners) are in the lead for development (e.g. cases from Belgium, Norway). The consequence of which is that tensions between actors can increase and might hamper site development, as the British case shows.

Considering the increase of complexity in land development processes (see also Hartmann & Spit, 2015), the question how the actors optimise their goals and how can we explain their behaviour, becomes more and more interesting. The next section will enlighten us in this point.

4 Evaluation: The Pipeline Effect

4.1 Running Time and Interdependency

The realisation of spatial developments takes a lot of time. The estimates run from five years to more than 20 years (e.g. the *Vinex* in the Netherlands). It speaks for itself that the larger the project or program, the longer it takes realising it. As the running time is that long, the contextual conditions for spatial developments are unstable by definition and the risks for the actors involved are enormous. Risk avoiding behaviour and minimisation of contextual risk factors can be considered to belong to the standard qualities of all actors. Their basic instincts tell them to protect their already running projects. The result of which is a conservative attitude of most of the actors involved in land developing processes.

During the long running time of land development projects, programs and/or processes all actors depend heavily upon each other (see the examples in the cases of Austria and Germany). Above and beyond that, considering that there is no blank starting point, the room to manoeuvre for actors can be considered small. This combination may even create some sort of trap for actors: a sort of path dependency may pop up, as the Norwegian case shows.

4.2 The Pipeline Effect

Elaborating on that, the importance of all limiting conditions, we believe that the starting position of each actor is by far the most important one. Especially when it concerns landownership (in the Belgian case: it might even create a 'landowners paradise'). However, it is the sum of all limiting conditions that is determining the little room to manoeuvre for actors. This is what we would like to call the "pipeline effect", which can be defined as is the result of all earlier actions, in combination with context conditions. The result of which is that there are little possibilities for actors to adapt to changing circumstances: all actors suffer more or less under the same 'tyranny of the Now' (see the Netherlands).

5 Reflection

5.1 Observations

Considering the role and behaviour of actors in the differing settings all over Europe, some observations seem unescapable. The first of which has to do with the former communist countries (e.g. Poland, Czechia etc.). In these countries, property rights seem to have become firmly rooted and therewith acknowledged in law. Although this turn towards a more market led economy might seem logical, its consequences for public power in development issues cannot be underestimated.

A second observation concerns the time frame. One might expect that land development, dealing with long term development processes by definition (five to 20 years, or more), is dominated by the consistency of public policy. This seems more or less logical, considering that private actors have to deal with market inconsistences day-in day-out and public actors have the advantage that they do not have to deal with every-day ups and downs of market developments. Although this might be true, development practice show that commercial developers are better equipped to deal with long term development goals (dominated by the desire for profit), than public authorities who suffer under other inconsistences such as politics and periodic elections. Continuity in development projects and processes seems a shared goal between actors. The examples in Europe show that the outcome of development processes illustrates the quality of this balancing act. A result such as "Mordor-like urban neighbourhood" (case from Poland) is an example of a process in which the balance was more important than its outcome. A third observation also has to do with the time frame of development processes. Actors and/or interests fluctuate in the duration of the process. This is obviously the case with both types of actors, public as well as private. The question however is, which type of actor suffers the most under these fluctuations. Looking at the development processes throughout Europe, changes in actors in the public domain seem to have more impact on the development processes than in the private domain, although there is no solid evidence to substantiate this.

The fourth and last observation is that public goals for development (e.g. social housing, sustainability aspects, urban green etc.) are always and everywhere under pressure (e.g. the Netherlands, Switzerland, Sweden etc.) when the development process continues. As all development processes constantly must adapt to changing circumstances, the costs of adaptation are systematically put on the account of these public goals (see e.g. case from Sweden). This seems to be a constant and irreversible trend.

5.2   Last Remark

All in all, the balance between public and private actors in development processes seem to be in favour of private parties. The fact that in some countries even public authorities initiate a 'publicly owned private partner' (e.g. Sweden, the Netherlands) can be seen as an illustration of the weak position of public actors in development processes.

References

Hartmann, T., & Spit, T. (2015). Dilemmas of involvement in land management—Comparing an active (Dutch) and a passive (German) approach. *Land Use Policy, 42,* 729–737.

Newman, P., & Thornley, A. (1996). *Urban planning in Europe.* Routledge.

Spit, T., & Zoete, P. (2016). PLANOLOGIE: Een wetenschappelijke introductie in de ruimtelijke ordening in Nederland. In Planning.

Witte, P., & Hartmann, T. (2022). *An introduction to spatial planning in the Netherlands.* Routledge.

Tejo Spit is a former Professor in Urban and Regional Planning at Utrecht University, The Netherlands. After his retirement in December 2021, at first he was active as a consultant engaged in complex spatial projects. Nowadays, his focus returned to research again as head of the Department of Spatial Planning and Quality of the Local Environment of the Netherlands Environment Assessment Agency in The Hague. This move mirrors an important characteristic in his career: combining the academic world with the more problem-oriented world of spatial planning.

A Reflection on Institutions of Land Policy

Sebastian Dembski

Abstract This chapter reflects on the institutional dimension of land policy in the twelve case studies. It reflects on the use of land policy instruments in different contexts and how the institution of private property is addressed. Whilst in some of the cases, there seems to be a real lack of suitable land policy instruments, in the majority of the cases the problem lies in the passive approach of the state in applying these. The solution is therefore not another planning reform, but a change of planning practice.

The twelve case studies of how land policy operates in each country provide an excellent snapshot of contemporary issues in relation to planning institutions, albeit some being as old as planning. The cases are different in location, scale and institutions, in particular the land policy instruments involved. Some focus on the role of formal plans allocating land for development (e.g. the case studies from Czechia, England, Norway), though the majority is concerned with land assembly and delivery. The following offer some reflections on different aspects of planning from an intuitional perspective, looking at how planning affected property rights both restricting and expanding them.

The role of private property rights is central to planning for sustainable urbanisation (Evers et al., 2024), where the ambition is to reduce land take in a context of housing need. There are two important issues at stake. The first relates to the question of excessive building land reserves, the second to the need to mobilise building land. The amount of building land reserves varies, but it is particularly the sites outside built-up areas that undermine the ambition to reduce net land take. Development rights have remained unused often for decades, but rezoning will not only limit the use but also result in a reduction of the land value which typically entails the right for compensation. The Belgian case study provides fascinating insights in the challenges of restricting development rights and in so doing undermine attempts to reduce net land take, as binding land use plans in the 1970s allocated an excessive amount of

S. Dembski (✉)
University of Liverpool, Liverpool, UK
e-mail: sebastian.dembski@liverpool.ac.uk

© The Author(s) 2025
T. Hartmann et al. (eds.), *Land Policies in Europe*,
https://doi.org/10.1007/978-3-031-83725-8_16

235

building land. The fact that any changes to property rights must be compensated seems to be primarily a political decision based on strong-rooted private property in a country where building your own house is considered the norm rather than a question of legality and varies strongly from neighbouring European countries (Lacoere et al., 2023). This is an example where the explanatory power of an instrumental rationality is limited.

Virtually all case studies highlight the importance of land as critical resource, yet the role of the state, usually the municipality, shows a great degree of variation. In the case studies from Austria, Finland, Sweden and Switzerland, public landownership was key in delivering the respective project, whereas in the cases studies in Czechia, England, Germany, France the state was more passive and at best providing incentives. The problem with a passive role of the state means that the state only benefits from the land value uplift if instruments to capture the land value uplift exist and are applied, and that development might stall due to landowners' unwillingness to sell or develop (e.g. the case studies from France, Germany). Research has shown the important role of the state as 'market maker' (Lord & O'Brien, 2017). Whilst the case studies are not necessarily representative for each country, some planning systems are known to have a more 'active' or 'passive' tradition of land policy (Hartmann & Spit, 2015). It would also be wrong to conclude that developers prefer a passive state. In the Dutch case study, developers are likely to voluntarily sell off their land to the municipality in exchange for serviced plots with development rights.

However, an active land policy highlights the importance of strong public justification and democratic control. In particular public landownership has many advantages (e.g. land assembly, land value capture, delivery), but it also poses an institutional dilemma as the state is creating its own development rights being proposer and approver of a development at the same time (Van der Krabben & Jacobs, 2013). But the issue applies more widely to close cooperations between developers and the state. In the Swiss case study, the special land use plan required approval by popular vote. The first plans were rejected, because the in the view of the majority of the voters the loss of open spaces outweighed the promise of not-for-profit housing. In the Swedish case study, despite municipal landownership, the large-scale development does not include any affordable housing, which is one of the central planning goals of the municipality. Delivering 'good housing for all' seems to be difficult in systems where all developers are treated equally and the municipality sells off the land to the highest bidder. The debates about who profits are not new, but in light of a renewed push for active land policy in the context of aiming to reduce land take and prioritise sites in built-up areas (Dembski et al., 2020), the grounding of planning policy in wider public norms is essential (cf. Salet, 2018).

Notwithstanding some caveats about active land policy (see Van der Krabben & Jacobs, 2013, for a fuller discussion), it is surprising how many countries lack effective policy instruments in particular for land assembly. In England's discretionary system, local planning authorities are almost entirely dependent on the private market with some of the common instruments assisting in land assembly, such as pre-emption rights and land readjustment, missing. Starved of cash, compulsory purchase is an option that is out of reach in most instances. In the French case study, it seems that mandatory land readjustment as in the Austrian case study could have solved the deadlock due to individual landowners unwilling to sell because of unrealistic expectations of land values. The Austrian case study shows that where there is mutual gain, landowners are happy to cooperate.

However, the lack of land policy instruments is only one part of the story. Also, there is widespread issues with the ability and willingness to apply land instruments. Even where there is a wide range of land policy instruments available, such as in Germany, municipalities often do not use these because of the risks involved or simply because of the costs compared to the benefits. In the German case study, where a city centre plot has been underutilised for decades, the reason for this did not seem to be a lack of land policy instruments. Also in the French case study, the municipality excelled through passivity and the heavy-handed instrument of a ZAC, which would enable land assembly and delivery in a concerted effort too resource intensive and complex for small-scale projects.

Whilst the discussion above related to restrictions of property rights, in many of the case studies, landowners are set to profit from a substantial expansion of property rights and public investment through planning (case studies from England, Finland, the Netherlands). The case study from Finland highlights the enormous upfront investments that are involved and, interestingly, that due to the existing practice of land value capture applied to private developments, would also lose out on income, which in this specific case meant that it was less profitable if standard percentages of land value capture had been applied. However, the case also shows that land value capture may reduce viability and in this case, there was no interest from the landowner to develop the side. That also highlights regional inequalities between prosperous and less prosperous real estate markets, with opportunities for land value capture being much lower or risking developments to stall (Lord et al., 2019).

Institutions are the cement of our societies and that includes but is not restricted to planning law. They establish mutual expectations in terms of behaviour and may involve sanctions. Institutions are therefore enabling and constraining at the same time (Scott, 2014). Institutional systems comprise of a regulative, normative and cultural-cognitive dimension each emphasising different mechanisms of enabling and constraining behaviour. Many institutionalists are particularly concerned with the regulatory aspects (cf. Scott, 2014), but institutions can be sustained by more than one or different pillars though time and they interact strongly with the other institutional elements (Scott, 2014: 62). Planning research on institutions, transaction costs and game theory approaches have proven important avenues in this regard (e.g. Samsura et al., 2010; Shahab & Viallon, 2019).

In planning practice, the importance of institutions beyond a purely instrumental view of achieving certain policy goals via regulation is often overlooked (Salet, 2018). Too often legislative changes have been announced with much fanfare only to discover that the underlying practices have changed very little (e.g. Buitelaar et al., 2011). Institutions include a regulatory dimension, but this is interwoven with broader public norms and symbolic systems, which together define the stability of institutions. Institutional change is poorly understood, and I will not attempt to provide an explanation. Research into critical junctures has been able to explain why radical change was possible in specific circumstances but suffice to say, it usually involves more than the approval of new legislation. But even if a new planning instrument is to be introduced successfully, there is still need for grounding in wider in society to properly function as structuring elements (Salet, 2018). Thus, a purely instrumental view of institutions will limit the ability to result in systemic change.

The institution of private property is critical in planning and a prime example of an institution that is not only formally regulated but also deeply embedded in social norms and symbolic systems. In his well-known essay, Krueckeberg (1995) argues that the planning and regulation of land use, that is the question where things belong, is less important that the question to whom things belong and in particular the definition, rights and distribution of property. The relation between planning and property is challenging as planning affects property rights (Blomley, 2017; Gerber et al., 2018) and therefore land values (e.g. Alterman, 2012). Whilst few property owners will complain if the value of their property increases due to planning and no investment of their own, invariably called betterment, unearned increment, etc., the opposite is the case when property rights are restricted, which is referred to as regulatory taking in a North American context to distinguish it from material taking as a result of compulsory purchase. It is here that planning leads to normative conflicts and where planning needs to be.

Whilst there are some pinch points in the way institutions of land policy function, frequent calls from planning professionals and politicians for planning reform need to be regarded with some caution as there are good reasons why planning law involves checks and balances. The aspirations of planners are conditioned by wider public norms (Salet, 2018) and even the reform or introduction of land policy instruments—itself a legal norm—is dependent on its justification and normative acceptance in a wider societal context. Institutions need to be validated in practice (Dembski & Salet, 2010). The balance between the aspirations of planners and institutions is fascinating as these are often not fully aligned. Reforms of planning law often remain without effect as real institutional change is lagging behind and institutional practices remain the same. The frequent changes of planning law, often with no effect on practices, has been called legislative ADHD (Teunissen in Buitelaar et al., 2013). The true work of planners lies in the application of the existing land policy instruments demonstrating a real need through changing practices, not numerous reforms of planning law that are usually too timid to be a game changer.

References

Alterman, R. (2012). Land use regulations and property values: the "windfalls capture" idea revisited. In: N. Brooks, K. Donaghy, & G.J. Knaap (Eds.), *The Oxford handbook of Urban economics and planning* (pp. 755–786). Oxford: Oxford University Press. https://doi.org/10.1093/oxfordhb/9780195380620.013.0034

Blomley, N. (2017). Land use, planning, and the "difficult character of property." *Planning Theory and Practice, 18*(3), 351–364. https://doi.org/10.1080/14649357.2016.1179336

Buitelaar, E., Galle, M., & Sorel, N. (2011). Plan-led planning systems in development-led practices: An empirical analysis into the (lack of) institutionalisation of planning law. *Environment and Planning A, 43*(4), 928–941. https://doi.org/10.1068/a43400

Buitelaar, E., Galle, M., & Salet, W. (2013). Third-party appeal rights and the regulatory state: Understanding the reduction of planning appeal options. *Land Use Policy, 35*, 312–317. https://doi.org/10.1016/j.landusepol.2013.05.011

Dembski, S., Hartmann, T., Hengstermann, H., & Dunning, R. (2020). Enhancing understanding of strategies of land policy for urban densification. *Town Planning Review, 90*(3), 209–216. https://doi.org/10.3828/tpr.2020.12

Dembski, S., & Salet, W. (2010). The transformative potential of institutions: How symbolic markers can institute new social meaning in changing cities. *Environment and Planning A, 42*(3), 611–625. https://doi.org/10.1068/a42184

Evers, D., Katurić, I., & Van der Wouden, R. (2024). *Urbanization in Europe: Past developments and pathways to a sustainable future.* Palgrave Macmillan. https://doi.org/10.1007/978-3-031-62261-8

Gerber, J.-D., Hartmann, T., & Hengstermann, A. (Eds.). (2018). *Instruments of land policy: Dealing with scarcity of land.* Abingdon.

Hartmann, T., & Spit, T. J. M. (2015). Dilemmas of involvement in land management—comparing an active (Dutch) and a passive (German) approach. *Land Use Policy, 42*, 729–737. https://doi.org/10.1016/j.landusepol.2014.10.004

Krueckeberg, D. A. (1995). The difficult character of property: To whom do things belong? *Journal of the American Planning Association, 61*(3), 301–309. https://doi.org/10.1080/01944369508975644

Lacoere, P., Hengstermann, A., Jehling, M., & Hartmann, T. (2023). Compensating downzoning. A comparative analysis of European compensation schemes in the light of net land neutrality. *Planning Theory and Practice, 24*(2), 190–206. https://doi.org/10.1080/14649357.2023.2190152

Lord, A., Burgess, G., Gu, Y., & Dunning, R. (2019). Virtuous or vicious circles? Exploring the behavioural connections between developer contributions and path dependence: Evidence from England. *Geoforum, 106*, 244–252. https://doi.org/10.1016/j.geoforum.2019.07.024

Lord, A., & O'Brien, P. (2017). What price planning? Reimagining planning as "market maker." *Planning Theory and Practice, 18*(2), 217–232. https://doi.org/10.1080/14649357.2017.1286369

Salet, W. (2018). *Public norms and aspirations: The Turn to Institutions in Action.* Routledge.

Samsura, A., Van der Krabben, E., & Van Deemen, A. (2010). A game theory approach to the analysis of land and property development processes. *Land Use Policy, 27*(2), 564–578. https://doi.org/10.1016/j.landusepol.2009.07.012

Scott, W. R. (2014). *Institutions and organizations: Ideas, Interests, and Identities* (4th ed.). Sage.

Shahab, S., & Viallon, F.-X. (2019). A transaction-cost analysis of Swiss land improvement syndicates. *Town Planning Review, 90*(5), 545–565. https://doi.org/10.3828/tpr.2019.34

Van der Krabben, E., & Jacobs, H. M. (2013). Public land development as a strategic tool for redevelopment: Reflections on the dutch experience. *Land Use Policy, 30*(1), 774–783. https://doi.org/10.1016/j.landusepol.2012.06.002

Sebastian Dembski is a Senior lecturer in Planning in the Department of Geography and Planning at the University of Liverpool, United Kingdom. He trained as a planner at TU Dortmund University, Germany, and did a PhD at the University of Amsterdam, Netherlands. He has written extensively on institutions from a comparative perspective.

Conclusions on Land Policies in Europe

Thomas Hartmann, Andreas Hengstermann, Mathias Jehling, Arthur Schindelegger, and Fabian Wenner

Abstract This concluding chapter presents four main conclusions based on the 12 cases and the reflections on land policy in European countries (AT, BE, CH, CZ, DE, ENG, FI, FR, NL, NO, PL, SE): Although land policy closely relates to spatial planning, it differs in two dimensions: (1) in its allocative and distributive perspective and (2) at strategic and operational levels of public policy. Based on this distinction four elements of land policy emerge: land use policy, land use planning, land tenure policy, and land management. Land policies comprise intentional decisions or activities led by public actors, sometimes with private sector cooperation. Land policy strategies stem from different interpretations of landowners' roles, which are often more decisive than specific instruments. Public actors conceptualise the relationship between property rights and land policy: To one end, landowners are key stakeholders; at the other end, they are partners in spatial projects. The chapter concludes that the international comparative analysis can deepen the understanding of land-use planning, property rights, and spatial development.

T. Hartmann (✉)
TU Dortmund University, Dortmund, Germany
e-mail: thomas.hartmann@tu-dortmund.de

A. Hengstermann
Norwegian University of Life Sciences, Ås, Norway
e-mail: andreas.hengstermann@nmbu.no

M. Jehling
Leibniz Institute of Ecological Urban and Regional Development, Dresden, Germany
e-mail: m.jehling@ioer.de

A. Schindelegger
BOKU University, Vienna, Austria
e-mail: arthur.schindelegger@boku.ac.at

F. Wenner
RheinMain University of Applied Sciences, Wiesbaden, Germany
e-mail: fabian.wenner@hs-rm.de

© The Author(s) 2025
T. Hartmann et al. (eds.), *Land Policies in Europe*,
https://doi.org/10.1007/978-3-031-83725-8_17

The examples presented in this volume demonstrate how land use and distribution are influenced by land policies, as explored through a variety of local case studies across Europe. Each case study begins by examining the country-specific perspective on land policy, understood as the interaction between (a) landowners—who typically pursue private interests, often with a focus on financial gains—and (b) public actors, particularly planning authorities, who aim to serve public interests, such as the provision of housing or the reduction of land take. While these case studies do not aim to offer a comprehensive representation of a country's entire land policy, they provide valuable insights into the specific mechanisms at work, illustrating how individual land policies are applied in local contexts.

The cases are structured around three key elements of land policy—resources, actors, and institutions—which were further elaborated upon in the introduction. These shared elements facilitate a cross-case reflection, explored through the reflections of Davy, Spit, and Dembski. Davy, in his reflection on resources, emphasises how prevailing norms and context shape the concept of land as a resource. By analysing the terminology used for land, he illustrates the differing interpretations and objectives of land policies across the cases. Spit's reflection on actors identifies a "pipeline effect", where actors in each case are constrained by earlier decisions, shaping the trajectory of the land policies in question. He also underscores the ongoing renegotiation and reinterpretation of what characterises the public and private spheres in land policy. Dembski's reflection on institutions focuses on property rights as a critical factor in promoting sustainable urbanisation and influencing actors' decisions. He also addresses questions of legitimacy and the role of land policy instruments, identifying these as common themes across the cases.

The interplay of resources, actors, and institutions in shaping spatial development serves as a foundation for developing a broader understanding of land policies in Europe. The comparative nature of this book supports four main findings: (1) a shared understanding of land policy, (2) land policy as an intentional public policy, (3) recognition of the strategic dimension of land policy beyond its individual instruments, and (4) the evolving role of landowners within land policy. These four elements will be elaborated upon in the subsequent four sections.

1 A Shared Understanding of Land Policy

The case studies in this book highlight that land policy is often closely related to, but distinct from, spatial planning. To foster a common understanding of land policy, it is important to clarify the difference between the two. Both terms—*land policy* and *(spatial) planning*—are used inconsistently across academia, practice, and national contexts (see, e.g. Nadin et al., 2018, 2024; Reimer et al., 2014; Sykes et al., 2023). This variation necessitates a clearer framework for understanding land policy. Based on the cases presented, we define land policy as encompassing strategic and operational public interventions in both how land may be used (allocation) and by whom it is used (distribution). Within this broader definition, we propose two dimensions of land policy (see Table 1):

1. The distinction between the *allocative* and *distributive* aspects of land property, and
2. The strategic and operational levels of public policy.

First, land policy addresses both the allocation and distribution of land (Gerber et al., 2018a, 2018b; Needham et al., 2018). Only when both of these aspects are combined can we truly talk about land policy (Hengstermann & Gerber, 2015), as all our cases have demonstrated. Policy interventions that focus on either allocation or distribution alone may be considered land-relevant policies, but they do not constitute a complete land policy. While some literature (e.g., Davy, 2005), introduces land value as a third aspect of land policy, we view land value as a *consequence* of the relationship between allocation and distribution, rather than independent aspect (Hengstermann & Gerber, 2015).

Second, land policy operates at both strategic and operational levels (Kötter, 2001; Krigsholm et al., 2022). Although these levels are not inherently scale-independent, in many of the cases presented here, the operational level tends to be associated with municipal or executive actions (Shahab et al., 2021), while the strategic level tends to operate at regional or national scales, aligned with legislative processes (Davy, 2012). Rarely are both strategic and operational levels, as well as the allocative and distributive aspects, governed by a single land law. More often, common land

Table 1 Strategies of land policy

Strategy of land policy	No explicit land policy	Regulative land policy	Cooperative land policy	Active land policy
Perception of landowners	Landowners are ignored	Landowners need to be regulated	Landowners as partners	Public landownership as solution
Planning and property	Planning without property	Planning against property	Planning with property	Planning by property
Use and disposal rights	No connection of use and disposal rights	Use rights shape disposal rights	Disposal rights shape use rights	Use rights and disposal rights aligned
Examples of cases	Bouleurs in the Greater Paris Area, France Burgwall 21 in Dortmund, Germany	Akkerstraat and Slotendries in Ghent, Belgium K Pazderkám in Prague, Czechia Wirral, UK	Rijnenburg in Utrecht, the Netherlands Mindemyren in Bergen, Norway Służewiec in Warsaw, Poland Norra Djurgårdsstaden in Stockholm, Sweden	Kaisers in Sölden, Austria Viererfeld in Bern, Switzerland Penttilänranta in Joensuu, Finland

		Public Policy	
		Strategic (legislative) abstract-general	Operational (executive) location-specific
Property rights in land	Land use (allocation)	Land Use Policy	Land Use Planning
	Land distribution	Land Tenure Policy	Land Management

Fig. 1 Components of land policy

policy components are dispersed across various laws, such as planning acts and civil codes, making it difficult to identify where specific components of land policy are regulated. In public policy theory, the strategic and operational levels can be linked to the stages of the public policy cycle: strategic planning aligns land use policies with land tenure policies, while operational planning implements land use through targeted interventions (Knoepfel 2007: 22). That align the allocation and distribution of land. In this sense, land policy encompasses public policy interventions that coordinate the allocation and distribution of land through strategic combinations of instruments.

The relationship between the two dimensions—property rights in land and public policy—can be mapped across four key components (see Fig. 1): land use policy, land use planning, land tenure policy, and land management.

- **Land Use Planning** involves preparing, designing, and controlling spatial development. This is often referred to as *urban and regional planning* in the U.S. terminology, or *town and country planning* in the U.K. In continental Europe, the term *spatial planning* is more prevalent, evolving since the 1980s as a broad encompassing term (Hengstermann, 2012; Nadin et al., 2024). It emphasises planning across various scales—from villages to regions or states—and integrates multiple sectors, such as transport and nature conservation. Spatial planning typically distinguishes between strategic and operational functions. The operational side, known as land use planning or zoning involves designating desired land uses and locations in specific areas. This process is location-specific and is largely 'property-blind', as it is intended to determine the best possible use of land without regard to landowner interests.
- **Land Use Policy** represents the strategic side of planning, which sets abstract, general goals for spatial development. The terms is often applied within the European Union (e.g., Richardson, 2000),while in national contexts, terms like 'planning policy' or issue specific policies such as housing policy or 'no-net-land take' policy are more common. While both land use policy and land use planning focus on the type and location of development (Hartmann & Spit, 2018), land use

policy outlines objectives at a broader, non-location-specific level and is similarly property blind.

- **Land Management** deals with the distributional aspects of spatial development. Land policy "commonly covers land development issues that are broader than land use" (Alterman, 1990: 16). Like planning, land management can be divided into strategic and operational levels. The operational, case-specific dimension is referred to by various terms across different countries, including 'land management' (DE), 'operational urbanism' (PL), or 'land development' (NL). In this context, there is significant overlap with geodesy, which takes an instrumental and managerial approach. Pure land management concerns ownership rights without consideration of land use planning, as seen in cadastres.
- **Land Tenure Policy** is a strategic approach that governs who has access to land and under what conditions. Although not explicitly used in our case studies, the concept is often used within the context of developing countries and the Global South. Institutions such as the World Bank, FAO, and the European Union refer to land tenure policy when discussing the redistribution of land access, particularly for rural and impoverished communities. In the European and urban context explored in this book, land tenure policy is less explicitly addressed, as the era of intense political debates over land redistribution has largely passed—despite significant redistribution processes still occurring (Bunkus & Theesfeld, 2018; Van Der Ploeg et al., 2015; Visser & Spoor, 2011). However, since cadastres are considered well-established and the development rights are secured, property distribution is often treated as an apolitical matter (Knoepfel et al., 2012).

By describing and distinguishing these four components, it becomes clear that land policy and planning are distinct elements. In this conclusion, we argue that land policy encompasses both the use and distribution aspects of the resource dimension as well as the operational and strategic aspects of the governance dimension, positioning it as the overarching concept.

The cases presented show that implementing spatial development objective is less effective when these four components are not aligned. For instance, the Bouleurs (France) case exemplifies how zoning and an infill development strategy prevented the municipality from achieving a no-net-land take goal, clashing with property owners' and developers' interest in urban expansion. Likewise, the cases in Germany and Poland illustrate what happens when planning adheres to a rationalistic approach, assuming that plans will be realised simply because they exist. This overlooks the fact that planning never acts in a vacuum—it must always account for landowners and their interests. Planning is inherently embedded in land markets, as every parcel of land is owned by either private, public, or commercial entities. Viewing planning as a component of land policy allows for a more critical analysis of spatial development and its associated objectives.

A conceptualisation of land policy components also highlights that land management can facilitate the achievement of land use planning goals and vice versa. Land use planning typically grants landowners usage rights without imposing obligations, leading to an increase of the economic value of the land, which landowners can potentially exploit (Kolocek & Hengstermann, 2020). Land management, on the other hand, provides policymakers with tools to ensure the implementation of land use plans. Often, these tools are embedded in domains complementary to planning law or based on private law. Strategic approaches to distribution can combine a diverse set of instruments to achieve overall spatial development goals, as exemplified by the combination of both dimensions of land policy in the successful implementation of affordable housing in Austria.

2 Framing Land Policy as Intentional Public Policy

Public policy can be defined as "a series of intentionally coherent decisions or activities taken or carried out by different public—and sometimes—private actors [...] with a view to resolving in a targeted manner a problem that is politically defined as collective in nature. This group of decisions and activities gives rise to formalised actions of a more or less restrictive nature that are often aimed at modifying the behaviour of social groups presumed to be at the root of, or able to solve, the collective problem to be resolved [...] in the interest of the social groups who suffer the negative effects of the problem in question" (Knoepfel, 2007: 22). Based on this definition, land policy can be considered a form of public policy, even though its specific may vary across the cases we presented.

As demonstrated in various European cases, land policies involve a series of intentional—ideally coherent—decisions or activities led by public actors, sometimes in cooperation with private actors, to address collective issues related to land use and allocation. These cases highlight that land policies address critical societal concerns such as housing affordability, land take, or densification, all of which are collective problems affecting communities and societies. While public authorities are the primarily drivers of land policies, private actors (e.g., developers, businesses, landowners) often play an essential role. This fits within the framework of public policy, where public interventions seek to influence the behaviour specific groups (in this case, landowners) to resolve or mitigate the initial political problem. When public objectives must be achieved on privately owned land, private stakeholders' interests become central—i.e. as Knoepfel (2007) noted, the behaviour of a social group is modified. So, land policies require the mobilisation of public policy resources (such as public landownership, zoning regulations, or financial incentives) and require institutional collaboration (e.g., between local and national governments, planning bodies, and environmental agencies), all key components of public policy. These intentional and coherent decisions or activities form a strategy, which is enacted through various land policy instruments. Therefore, understanding land policies also requires understanding these instruments and how they are strategically combined.

3   Strategic Use of Land Policy Instruments

The implementation of land policies occurs through instruments applied by various actors (Gerber et al., 2018a, 2018b). In public policy, these instruments can be described as "formalised actions" (Knoepfel, 2007) or institutions (Ostrom, 2007), as they provide actors with the agency to achieve strategic goals. It is assumed that actors follow specific strategies when applying these instruments (Hood, 1983). Effective land policy strategies should aim harmonise land use and distribution, including both use and disposal rights. In the cases presented, the public sector approaches these challenges in different ways (see below). These strategies also differ from those of other actors, primarily municipalities, within the same country. Accordingly, it is inaccurate to speak of a singular, cohesive land policy in Europe—this is why the title of our book is not 'European Land Policy' but 'Land Policies in Europe'.

Examining the cases presented, highlights the importance of strategy in land policy implementation. The application of land policy instruments by actors depends on their respective strategies. At the individual project level, established practices and political objectives together provide actors with the tools to either advance or obstruct spatial development projects (Shahab et al., 2021). Each case represents a specific, contextual approach to combine instruments that address both property rights and public policy to resolve conflicts over urban land use.

Several general land policy strategies emerge across the cases. These strategies illustrate how public actors approach land ownership and how they align land use and distribution through the strategic use of instruments from public or private law (Gerber et al., 2018), addressing both property rights and public policy. The case of *Viererfeld*, Switzerland demonstrates how a public planning authority can adopt a purely private law approach, acquiring land while simultaneously setting up land use regulations. In contrast, the *Wirral* case in England shows how a private actor can attempt to align land use planning with private development interests by deviating from established planning on a project basis. The case of Ghent, Belgium, reveals how the institution of property rights stemming from former land use planning decisions can create path dependency in municipal planning, with previous planning decisions limiting changes to municipal land use.

These comparative insights underscore that differing land policy strategies are rooted in varying interpretations of the role of landowners in land policy. These interpretations are often more decisive than the specific land policy instruments used. Remarkably, the cases do not showcase a wide range of diverse public policy instruments, such as pre-emption rights, expropriation, or other regulatory measures. Contrary to the claims made by Hengstermann et al. (2023), it is difficult to argue that the cases presented in this book are characterised by innovative instruments. Instead, they demonstrate how conventional instruments are strategically combined.

4   Public Authority—Landowner Interaction: Towards Common Strategies of Land Policies

The cases presented illustrate the challenges of aligning public and private interests when conflicts arise from different ideas on the level of utilisation and disposal rights. To what extent are the operational elements of land use planning and land management in line? Effective land use planning requires that planners do no not underestimate the importance of aligning plans with landowner's expectations and perspectives. Plans only hold practical value when they not only reflect planning policy objectives but also account for the landowner's interests or when ownership is adapted through mechanisms such as purchase, land readjustment, pre-emption or, in theory, expropriation.

Land policy must balance the differing interests of landowners and planning authorities in spatial development. In doing so, spatial planners, whether implicitly or explicitly, interpret the role of the state and property rights by public authorities differently (Gerber, Hengstermann, & Viallon, 2018). In the introduction to this volume, we noted that "it seems that oftentimes planning practitioners perceive landowners merely as an obstruction to implementing spatial plans" (Hartmann et al., this volume). However, the cases reveal a variety of ways in which planning authorities perceive landowners.

These cases reveal a spectrum of how public actors conceptualise the relationship between property rights and land policy (see table D.1.2). At one end, there is the tendency to overlook landowners as key stakeholders in the planning process—which sometimes leads to almost fatalistic behaviour on the part of planning authorities. At the other end, landowners are seen as partners in the collaborative development of spatial projects. For instance, the Swiss *Viererfeld* case illustrates a more entrepreneurial approach, where public authorities themselves become landowners, thereby avoiding complex negotiations with private landowners or regulatory constraints on property use. Only a few cases—such as the *Burgwall* case in Germany and the *Bouleaurs* case in France—demonstrate attempts by public authorities to govern land property solely through regulations, with very limited success. This appears to be the least appealing and least effective throughout the cases.

The way land policy relates planning to private property highlights how use and disposal rights are aligned in the cases. Use rights refer to the entitlement to access and utilise land for specific purposes, while disposal rights give the landowner the authority to transfer, sell, or relinquish ownership or control of the land. Unlike use rights, which focus on how the land is utilised, disposal rights determine who has the power to make long-term decisions about the land's ownership and future. Across the cases, four types of land policy strategies can be identified (see Table 1).

The four strategies of land policy are represented in the different cases as follows:

- **No explicit land policy**: In some cases, landowners are framed as obstructive or adversarial to public authorities rather than as key stakeholders in implementing planning objectives. In these cases, landowners are viewed as irrelevant to planning. This is illustrated by the Burgwall case in Germany, where the municipality refrained from intervening in the owner's property rights. Similarly, the Bouleurs case in France exemplifies this lack of interaction, as the planning authority was unable to steer spatial development to the desired location and had to accept development in a less favourable location. Planners in these situations seem to accept that spatial planning merely establishes primary conditions for land use, without wishing to intervene in landowners' decisions regarding the timing and extent of development. This perspective aligns with the interpretation that property rights provide "individual liberty, political stability, and economic prosperity" (Ellickson, 1993, p. 1317). Consequently, land policy appears somewhat detached from land use planning, with property considerations not actively addressed in plans.

- **Regulative land policy**: In some cases, landowners are acknowledged as important stakeholders in realising spatial development projects, but public authorities attempt to influence their behaviour through regulations. Various instruments and strategies are employed to effect landowners' use rights. The cases from Czechia, Belgium, and, to some extent, England, align with this land policy strategy. Here, the approach to land markets and property rights is more passive, with public authorities not actively using property in land as a strategic tool to achieve planning objectives.
- **Cooperative land policy**: In other instances, a cooperative notion prevails. Public authorities conceive landowners as partners to realise the public policy objective. The Polish case, the Dutch Rijnenburg case, as well as examples from Norway and Sweden illustrate this strategy in practice. Often, spatial development contracts formed by specific development organisations operate as public–private partnerships. In such cases, disposal rights in land shape use rights and plans. They might be initiated or co-initiated by private developers or landowners, prompting a reaction by public authority to respond to these initiatives.
- **Active land policy**: This land policy strategy involves public authorities actively becoming landowners. In this approach, disposal rights are concentrated in the hands of the public authorities—examples include cases from Switzerland, the Finland, and Austria. Here, landownership is considered the solution to realising planning objectives.

This categorisation of the four land policy strategies is not absolute; however, across the diversity of European cases, the categorisation can help distinguish different approaches to land policy.

5   The Added Value of Comparative Land Policy

This book focuses on land policies in Europe while presenting a selection of cases and countries. Nonetheless, the comparative approach demonstrates the potential for mutual learning about land policy across different nations. This endeavour is undoubtedly challenging, but it is valuable to understand land policies in diverse contexts.

The case studies presented illustrate a variety of spatial settings in which land policies are applied across European countries. They also highlight the heterogeneity of political goals, providing insight into the spatial context, the actors involved, and the institutions at play. This comprehensive perspective enables us to grasp practices beyond national boundaries, fostering further comparative learning and knowledge building in land policy development.

The goal should not be to replicate strategies or instruments from other countries in a legal transplant manner (Watson, 1974). Instead, we aim to reflect critically on the land policies in practice and draw inspiration (Zimmermann et al., 2023). There are various lessons to learn for the diverse readerships of this book:

Policymakers and legislators face unique challenges. They act on behalf of the public interest and are tasked with addressing specific issues through legislation. In our field, this often entails introducing new instruments. While this instrumentalism is understandable, some of the editors and authors of this book have also encountered this tendency (Gerber et al., 2018a, 2018b). However, this new volume, as a continuation of the ongoing debate, emphasises that laws and their instruments cannot be viewed in isolation. For them to be effective, they must be embedded within strategies, enacted by actors, and related resource and their associated challenges. We therefore encourage policy-makers and legislators to continue to advocate actively for our societies and to adopt a holistic understand of the cases presented.

The reality is much more complex than can be captured in this book. The cases we have included represent only a sample, and any practitioner could rightly point out that numerous additional factors influence these cases beyond what we have discussed. The editors and authors of this book have the privilege and disadvantage of occupying an ivory tower, which allows us to understand reality only in a simplified manner. Therefore, practitioners should focus less on critiquing the omissions in the chapter about their own country and instead seek inspiration from the chapters on other countries.

As academics, our primarily interest lies in developing our students' skills, as they will ultimately influence the use and distribution of land in the future. We encourage them to remain curious—and critical. Everything they learn in our lectures about system operations should be questioned. This book effectively illustrates that land use planning and property rights are well-established in individual countries—but there are also viable alternatives. Exploring approaches abroad reveals a wealth of alternative strategies. Thus, nothing should be taken for granted. For a deeper understanding of land-use planning, property rights, and spatial development, we therefore invite you, the reader, to go abroad and experience these concepts first-hand.

References

Alterman, R. (1990). *Private supply of public services: Evaluation of real estate exactions, linkage, and alternative land policies.* New York University Press.

Bunkus, R., & Theesfeld, I. (2018). Land grabbing in Europe? Socio-cultural externalities of large-scale land acquisitions in East Germany. *Land, 7*(3), 98.

Davy, B. (2005). Bodenpolitik. In E. H. Ritter (Ed.), *Handwörterbuch der Raumordnung* (4, neu bearb. Auflage), pp. 117–130. ARL.

Davy, B. (2012). *Land policy: A German perspective on planning and property.* Ashgate.

Ellickson, R. (1993). Property in Land. *Yale Law Review* (102), 1315–140.

Gerber, J.-D., Hengstermann, A., & Viallon, F.-X. (2018). Land policy: How to deal with scarcity of land. In J.-D. Gerber, T. Hartmann, & A. Hengstermann (Eds.), *Instruments of land policy: dealing with scarcity of land* (pp. 8–26). Routledge.

Gerber, J. D., Hartmann, T., & Hengstermann, A. (Eds.). (2018). *Instruments of Land Policy: Dealing with Scarcity of Land.* Routledge.

Hartmann, T., & Spit, T. (2018). Editorial: Dynamics of land policies—triggers and implications. *Land Use Policy, 77*, 775–777. https://doi.org/10.1016/j.landusepol.2018.01.042

Hengstermann, A. (2012). Geschichte der Raumplanung auf Europäischer Ebene.*RaumPlanung* 165(6), pp. 51–55. Dortmund: Informationskreis für Raumplanung IfR.

Hengstermann, A., & Gerber, J. D. (2015). Aktive Bodenpolitik—Eine Auseinandersetzung vor dem Hintergrund der Revision des eidgenössischen Raumplanungsgesetzes. *Flächenmanagement und Bodenordnung, 2015*(6), 241–250.

Hengstermann, A., Wenner, F., Jehling, M., & Hartmann, T. (2023). Innovative land policies in Europe. *Raumforschung Und Raumordnung, 81*(6), 575–578. https://doi.org/10.14512/rur.2246

Hood, C. (1983). *The tools of government.* Macmillan.

Knoepfel, P. (2007). *Public policy analysis.* Policy Press.

Knoepfel, P., Csikos, P., Gerber, J. D., & Nahrath, S. (2012). Transformation der Rolle des Staates und der Grundeigentümer in städtischen Raumentwicklungsprozessen im Lichte der nachhaltigen Entwicklung. *Politische Vierteljahresschrift*, 414–443.

Kolocek, M., & Hengstermann, A. (2020). The myth of responsiveness. Discourse analysis on the indirect effectiveness of building orders and planning-law expropriations. *Raumforschung und Raumordnung\ Spatial Research and Planning, 78*(6), 559–573.

Kötter, T. (2001). Flächenmanagement—zum Stand der Theoriediskussion. In *fub – Flächenmanagement und Bodenordnung, Heft 4/2001* (pp. 145–166).

Krigsholm, P., Puustinen, T., & Falkenbach, H. (2022). Understanding variation in municipal land policy strategies: An empirical typology. *Cities, 126*, 103710.

Nadin, V., Fernandez Maldonado, A. M., Zonneveld, W. A. M., Stead, D., Dabrowski, M. M., Piskorek, K. I., Sarkar, A., Schmitt, P., Smas, L., & Cotella, G. (2018). *COMPASS Comparative Analysis of Territorial Governance and Spatial Planning Systems in Europe. Applied Research 2016–2018: Final Report.* ESPON & TU Delft.

Nadin, V., Cotella, G., & Schmitt, P. (2024). Spatial planning systems: a European perspective. In *Spatial Planning Systems in Europe* (pp. 2–27). Edward Elgar Publishing.

Needham, B., Buitelaar, E., & Hartmann, T. (2018). *Planning, law and economics: The rules we make for using land* (2nd ed.). Routledge.

Ostrom, E. (2007). Institutional rational choice—an assessment of the institutional analysis and development framework. In: P. Sabatier (Ed.), *Theories of the Policy Process* (pp. 21–64). Westview Press.

Reimer, M., Getimēs, P., & Blotevogel, H. H. (Eds.). (2014). Spatial planning systems and practices in Europe: A comparative perspective on continuity and changes (1st ed.). Routledge, Akademie für Raumforschung und Landesplanung.

Richardson, T. (2000). Discourses of rurality in EU spatial policy: The European spatial development perspective. *Sociologia Ruralis, 40*(1), 53–71.

Shahab, S., Hartmann, T., & Jonkman, A. (2021). Strategies of municipal land policies: Housing development in Germany, Belgium, and Netherlands. *European Planning Studies, 29*(6), 1132–1150.

Sykes, O., Shaw, D., & Webb, B. (2023). *International Planning Studies: An Introduction.* Springer Nature.

Van Der Ploeg, J. D., Franco, J. C., & Borras, S. M., Jr. (2015). Land concentration and land grabbing in Europe: A preliminary analysis. *Canadian Journal of Development Studies/revue Canadienne D'études Du Développement, 36*(2), 147–162.

Visser, O., & Spoor, M. (2011). Land grabbing in post-Soviet Eurasia: The world's largest agricultural land reserves at stake. *The Journal of Peasant Studies, 38*(2), 299–323.

Watson, A. (1974). *Legal transplants—an approach to comparative law.* The University Press of Virginia.

Zimmermann, K., Diller, C., & Othengrafen, F. (2023). Planungssysteme vergleichen—aber wie? *disP – The Planning Review, 59*(2), 38–52. https://doi.org/10.1080/02513625.2023.2257487

Thomas Hartmann is the chair of land policy and land management at the School of Spatial Planning, TU Dortmund University, Germany. His research focuses on strategies of municipal land policy, and the relation of flood risk management and property rights. He is the former president of the International Academic Association on Planning, Law, and Property Rights.

Andreas Hengstermann is an Associate Professor with the Department of Urban and Regional Planning (BYREG) at the Norwegian University of Life Sciences (NMBU). He holds a Ph.D. in Geography from the Institute of Geography of the University of Bern (CH). He was educated as a planner (Dipl.-Ing. Raumplanung) at the TU Dortmund University (DE) and the Universidad de Huelva (ES). Furthermore, he did postgraduate studies in public law (DAS Law). His primary research interest lies in understanding the essential role of property rights in shaping spatial development. His research includes comparative studies across different geographical contexts, political systems, and legislations. He current serves as Vice-President International Academic Association on Planning, Law, and Property Rights.

Mathias Jehling is senior researcher at Leibniz Institute of Ecological Urban and Regional Development (IOER) in Dresden, Germany, where he leads the research group on "Urban Structure and Policy". His focus is on geographic information in the planning context. He works and teaches (Technical University of Dresden) on urban form and institutionalist approaches to planning and land policies. He obtained his PhD at Karlsruhe Institute of Technology.

Arthur Schindelegger is a postdoc research fellow at University of Natural Resources and Life Sciences, Vienna (BOKU University), Institute of Landscape Planning. He has studied spatial planning at TU Wien and KTH Stockholm and has worked in the private sector for several years. His research is centred around questions on land policy, planning law and the integration of natural hazards and climate change adaptation in planning processes and procedures. He is also a consultant with the World Bank on resilient urban development.

Fabian Wenner is professor for sustainable urban and transport planning at RheinMain University of Applied Sciences in Wiesbaden, Germany. His research focuses on integrated transport and settlement planning through accessibility, instruments of land policy for sustainable urban development, and digitalisation in urban planning.